Lecture-Tutorials

FOR INTRODUCTORY ASTRONOMY

THIRD EDITION

Edward E. Prather
University of Arizona

Timothy F. Slater
University of Wyoming

Jeffrey P. Adams
Millersville University

Gina Brissenden
University of Arizona

with contributions by

Jack A. Dostal
Wake Forest University

Colin S. Wallace
University of Arizona

PEARSON

Boston   Columbus   Indianapolis   New York   San Francisco   Upper Saddle River
Amsterdam   Cape Town   Dubai   London   Madrid   Milan   Munich   Paris   Montreal   Toronto
Delhi   Mexico City   São Paulo   Sydney   Hong Kong   Seoul   Singapore   Taipei   Tokyo

Publisher: Jim Smith
Executive Editor: Nancy Whilton
Director of Development: Laura Kenney
Project Editor: Tema Goodwin
Director of Marketing: Christy Lesko
Marketing Manager: Will Moore
Managing Editor: Corinne Benson

Production Project Manager: Mary O'Connell
Production Service and Composition:
Thistle Hill Publishing Services, LLC and
  Cenveo Publisher Services/Nesbitt Graphics, Inc.
Art House: Rolin Graphics
Cover Design: Mark Ong
Text and Cover Printer: LSC Communications

Cover Image: SOHO (ESA & NASA)

Inside Back Cover Image: NASA and The Hubble Heritage Team (STScI/AURA)

Photos courtesy of: 81/85 UCO/Lick Observatory; 103 Palomar Observatory/California Institute of Technology; 135 European Southern Observatory/Photo Researchers, Inc.; 139 NASA/Keck Observatory; 145 European Southern Observatory/Photo Researchers, Inc.; 150 Jason T. Ware/Photo Researchers, Inc.

39 2020

PEARSON

ISBN 10: 0-321-82046-0
ISBN 13: 978-0-321-82046-4

Table of Contents

Instructor's Preface

Each year, over 250,000 students take introductory astronomy—hereafter referred to as Astro 101; the majority of these students are non-science majors. Most are taking Astro 101 to fulfill a university science requirement, and many approach science with some mix of fear and disinterest. The traditional approach to help students learn has been to emphasize creative and engaging lectures, taking full advantage of both demonstrations and awe-inspiring astronomical images. However, what a growing body of evidence in astronomy and physics education research has been demonstrating is that even the most popular and engaging lectures do not engender the depth of learning for which faculty appropriately aim. Rigorous research into student learning tells us that one critical factor missing from the traditional lecture-based classrooms is the ability to intellectually engage students in collaborative learning environments where they construct their own understanding while working through inquiry-based activities. This is best expressed in the mantra: "It's not what the teacher does that matters; rather it's what the students do that matters."

Lecture-Tutorials for Introductory Astronomy has been developed in response to the demand from astronomy instructors for easily implemented student activities for integration into existing course structures. Rather than asking faculty—and students—to convert to an entirely new course structure, our approach in developing *Lecture-Tutorials* was to create classroom-ready materials to augment traditional lectures. Any of the activities in this manual can be inserted at the end of lecture presentations, and because of the education research program that led to the activities' development, we are confident in asserting that the activities will lead to deeper and more complete student understanding of the concepts addressed.

Each *Lecture-Tutorial* presents a structured series of questions designed to confront and resolve student difficulties with a particular topic. Confronting difficulties often means answering questions incorrectly; this is expected. When this happens, the activities make use of additional questions or situated student debates designed to help a student understand where her or his reasoning went wrong and to develop a more thorough understanding as a result. Therefore, while completing the activities, students are encouraged to focus more on their reasoning and less on trying to guess an expected answer. The activities are meant to be completed by students working in pairs who "talk out" their answers and reasoning with each other to make their thinking explicit.

At the conclusion of each *Lecture-Tutorial,* instructors are encouraged to engage their class in a brief discussion about the particularly difficult concepts in the activity. The online *Instructor's Guide*[1] also provides "post-tutorial" questions that can be used to gauge the effectiveness of the *Lecture-Tutorial* before moving on to new material.

Unique to this third edition of *Lecture-Tutorials for Introductory Astronomy* are six new activities focusing on topics not found in the second edition. These all-new activities were specifically chosen to fill gaps from the second edition regarding the most common topics taught in an Astro 101 course. As a result, there are now new activities that focus on the greenhouse effect, dark matter, the Big Bang, and the expansion and evolution of the universe. It's now possible to complement your instruction with the "Greenhouse Effect," "Dark Matter," "Making Sense of the Universe and Expansion," "Hubble's Law," "Expansion, Lookback Times, and Distances," and "The Big Bang" *Lecture-Tutorials.* These new activities

---

[1]Instructors can go to http://www.pearsonhighered.com for an online *Instructor's Guide* that gives detailed information on classroom implementation as well as evidence of the efficacy of specific activities.

have been through the same rigorous development cycle that was used to create the highly successful activities of the first and second editions.

In addition, several changes have been made to your favorite activities from the first and second editions of *Lecture-Tutorials.* Over the last several years, we have performed continuous and systematic research to uncover places where students struggle with the wording of questions or scenarios presented in the *Lecture-Tutorial* activities. As a result, many activities from the first and second editions have been notably changed for the third edition. In particular, the diagrams, graphs, and artwork have been significantly improved to help students make hard-to-visualize and perceptually complex ideas more approachable and easier to comprehend.

Acknowledgments

Lecture-Tutorials for Introductory Astronomy was developed with generous support from the National Science Foundation (#0715517, #9952232, #9907755), the Jet Propulsion Laboratory's NASA Exoplanet Exploration Program, the Spitzer Education and Public Outreach Programs, Montana State University, the Conceptual Astronomy and Physics Education Research (CAPER) team, the University of Arizona, and the Center for Astronomy Education (CAE). Numerous individuals contributed to this project through critical assessment and the national field-testing of the materials. These individuals include Ingrid Balsa, Chija Skala Bauer, Erik Brogt, Tom Brown, Dave Bruning, Sébastien Cormier, Erin Dokter, Doug Duncan, Thomas Fleming, Seth Hornstein, Beth Hufnagel, John Keller, Janet Landato, Dan Loranz, Ed Murphy, Erika Offerdahl, and Larry Watson. Particularly noteworthy were the extensive reviews and suggestions provided by Janelle Bailey, Lauren Jones, Steve Shawl, and Alex Storrs, which continually kept us on our toes. In addition, we must thank Nancy Whilton and Tema Goodwin, who helped us with day-to-day publication issues. Most importantly, we wish to express our appreciation to all of the students who patiently endured early versions of these tutorials and unselfishly provided extensive feedback.

Note to the Student

Welcome to the study of astronomy! You are about to embark on a grand study of the cosmos. To help you better understand the topics of your course we have created this series of activities called *Lecture-Tutorials.* In each activity, you are asked a short series of questions that will require you to work in collaboration with your classmates to help you learn important and difficult concepts in astronomy. For every question in these activities, it is important that you write out a detailed answer. This is critical because you will certainly be using these materials to study for exams. It is also important because part of the learning process is being able to express complex ideas in writing.

We strongly encourage you to actively engage in completing these activities in collaboration with another student. The process of deciphering the questions and negotiating a common language to write your answers will help you understand the concepts more deeply. Specifically, the *Lecture-Tutorials* are designed to give you a starting point to think carefully and talk with others about concepts in astronomy. Above all, have fun exploring astronomy!

—Ed Prather, Tim Slater, Jeff Adams, and Gina Brissenden

In the celestial sphere model, Earth is stationary and the stars are carried on a sphere that rotates about an axis that points at the North Star. In Figure 1 below, two stars, A and B, are each shown at four different positions (1, 2, 3, and 4) through which each star will pass during the course of one revolution of the celestial sphere. In addition, your location on Earth in the Northern Hemisphere and the portion of the celestial sphere that is above your horizon are shown.

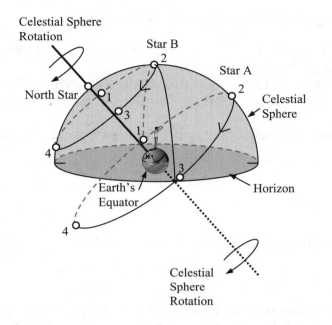

Figure 1

1) Is the horizon shown a real physical horizon or an imaginary plane that extends from your observing location on Earth out to the stars?

2) Can the observer shown see a star when it is located below the horizon? Why or why not?

3) Is either Star A or B ever in an unobservable position? If so, which position(s)?

4) When a star travels from a position below the observer's horizon to a position above the observer's horizon, is that star rising or setting?

5) When a star travels from a position above the observer's horizon to a position below the observer's horizon, is that star rising or setting?

6) Star A is just visible above your eastern horizon at Position 1. At which of the numbered positions is it just visible above your western horizon?

7) At which position(s), if any, does Star B rise and set?

8) Two students are discussing their answers to Question 7.

   Student 1: *Locations B1 and B3 are on my horizon because they are rising and setting just like A1 and A3.*

   Student 2: *Figure 1 shows that Star B is as low as it will get when it is just above the northern horizon at B4. So Star B never goes below the horizon.*

   Do you agree or disagree with either or both of the students? Explain your reasoning.

9) Label the directions north, south, east, and west on Figure 1. Check your answer with another group.

10) For each indicated position, describe where in the sky you must look to see the star at that time. Each description requires two pieces of information: the direction you must face (north, northeast, east, etc.) and how far above the horizon you must look (low, high, or directly overhead). If you cannot see the star, state that explicitly. The descriptions for four positions are given as examples.

   a) A1: *east, low*

   b) A2:

   c) A3:

   d) A4:

   e) North Star: *north, high*

   f) B1:

   g) B2: *directly overhead*

   h) B3: *northwest, high*

   i) B4:

   Check your answers with a nearby group and resolve any inconsistencies.

11) Does Star B ever set?

Part I: Looking North

For this activity, imagine you are the observer shown on Earth in the Northern Hemisphere and that the time is 6 P.M. Looking north, the sky will appear as shown in Figure 1. The positions and motions of the star in Figure 1 can be understood by imagining yourself as the observer at the center of the celestial sphere as shown in Figure 2. In the celestial sphere model, Earth is stationary and the stars are carried on a sphere that rotates about an axis that points at the North Star. Note that only the portion of the celestial sphere that is above your horizon is shown.

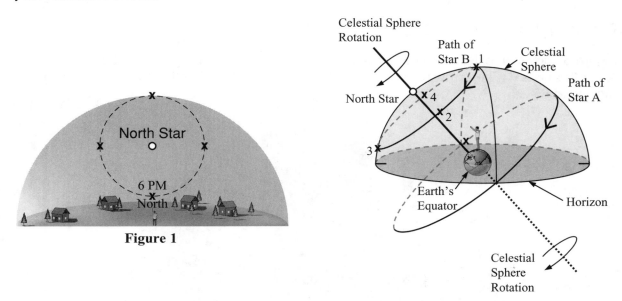

Figure 1

Figure 2

The **x**'s in both figures represent four of the positions through which Star B will pass during the course of one revolution of the celestial sphere. Ignore Star A until Question 6.

1) Note in Figure 1 that the position of Star B at 6 P.M. has been identified for you. Circle the numbered position (1, 2, 3, or 4) in <u>Figure 2</u> that corresponds to the identified location of Star B at 6 P.M. provided in Figure 1.

2) The rotation of the celestial sphere carries Star B around so that it returns to the same position at about 6 P.M. the next evening. Label each of the **x**'s in both figures with the approximate time at which Star B will arrive (e.g., the location you circled in Question 1 will be labeled "6 P.M.").

3) Using Figure 2, describe the direction you have to look to see Star B at 6 A.M.

4) The position directly overhead is called the **zenith**. Label the direction of the zenith on Figure 2. How does the direction of the zenith compare to the direction that you identified in Question 3?

5) In Figure 1, the path that Star B follows is shown with a dashed line. Draw a small arrowhead on the path to represent the direction Star B would be moving at the instant it is at each of the four locations marked with an **x**. Check your answers with a nearby group.

6) Using Figure 2, describe in words where you would look to see Star A when it is halfway between rising and setting.

Part II: Looking East

Figure 3 shows an extended view along the eastern horizon showing the positions of Stars A and B at 6 P.M. The arrow shown is provided to indicate the direction that Star B will be moving at 6 P.M.

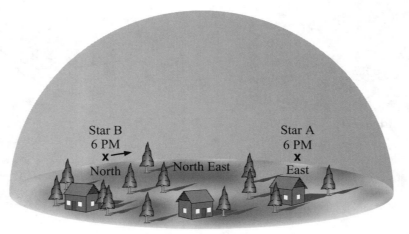

Figure 3

7) Recall that in Question 6, you found that Star A ends up high in the southern sky when it is halfway between rising and setting (and therefore never passes through your zenith). Draw a straight arrow at the **x** in the east in Figure 3 (the position of Star A at 6 P.M.) to indicate the direction Star A moves as it rises. Studying Figure 2 can also help clarify your answer.

8) Two students are discussing the direction of motion of a star rising directly in the east.

   Student 1: *Stars move east to west so any star rising directly in the east must be moving straight up so that it can end up in the west. If the arrow were angled, the star would not set in the west.*

   Student 2: *I disagree. From Figure 2, the path of Star A starts in the east, then it moves high in the southern sky yet still sets in the west. To do this it has to move toward the south as it rises so I drew my arrow angled up and to the right.*

   Do you agree or disagree with either or both of the students? Explain your reasoning.

9) Imagine you could see Star B at noon. Fifteen minutes later, in what direction will Star B have moved? Explain your reasoning.

10) Consider the student comment below.

Student: *The amount of time that all stars are above the horizon is 12 hours because it takes 12 hours for a star to rise in the east and then set in the west.*

Do you agree or disagree with the student? Explain your reasoning.

Consider the situation shown below in which the Sun and a group of constellations are shown at sunrise, Figure 4, and then shown again 8 hours later, Figure 5.

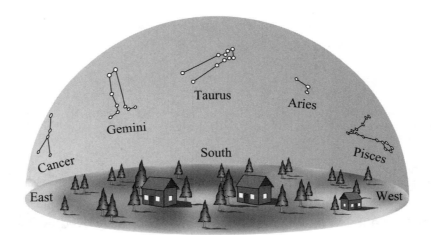

Figure 4

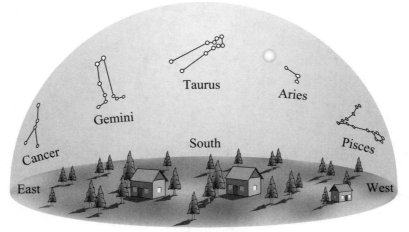

Figure 5

11) Consider the following debate between two students regarding the motion of the Sun and constellations shown in Figures 4 and 5.

Student 1: *We know the Sun rises in the east and moves through the southern part of the sky and then sets in the west. Eight hours after sunrise, it makes sense that the Sun will have moved from being on the eastern horizon near the constellation Cancer to being located high in the southwestern sky near the constellation Aries.*

Student 2: *You're forgetting that some stars and constellations also move from the east through the southern sky and to the west just like the Sun. So, the Sun will still be near Cancer eight hours later. So Figure 5 is drawn incorrectly. It should show that the constellations have all moved like the Sun, so Cancer would also be located high in the southwestern sky, with the Sun, eight hours later.*

Do you agree or disagree with either or both of the students? Explain your reasoning. Check your answers with another group.

12) In Question 11, we found that Figure 5 was drawn incorrectly. Redraw Figure 5 on the figure below by sketching the approximate location of any constellations from Figure 5 that would still be visible.

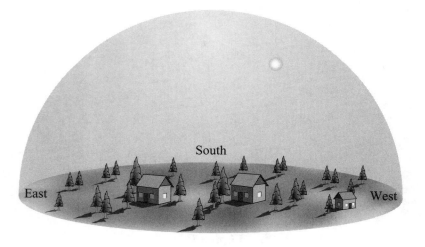

Part I: Monthly Differences

Figure 1 shows a Sun-centered, or heliocentric, perspective view of the Earth–Sun system indicating the direction of both the daily rotation of Earth about its own axis and its annual orbit about the Sun. You are the observer shown in Figure 1, located on Earth in the Northern Hemisphere while facing south.

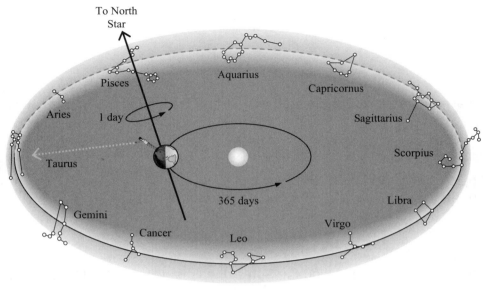

Figure 1

Figure 2 shows a horizon view of what you would see when facing south on this night at the same time as shown in Figure 1.

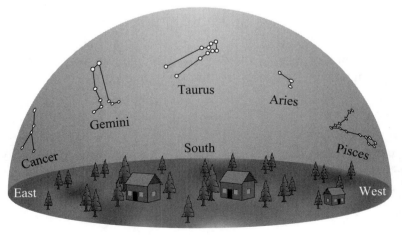

Figure 2

1) Which labeled constellation do you see highest in the southern sky?

2) For the time shown, which constellation is just to the east (i.e., to your left when you are facing south) and which constellation is just to the west (i.e., to your right when you are facing south) of the highest constellation at this instant?

   east: west:

3) Noting that you are exactly on the opposite side of Earth from the Sun, what time is it?

4) In six hours, will the observer be able to see the Sun? If not, why not? If so, in what direction (north, south, east, or west) would you look to see the Sun?

5) Which constellation will be behind the Sun at the time described in Question 4?

6) When it is noon for the observer, which constellation will be behind the Sun?

7) One month later, Earth will have moved one-twelfth of the way around the Sun. You are again facing south while observing at midnight. Which constellation will now be highest in the southern sky?

8) Do you have to look east or west of the highest constellation that you see now to see the constellation that was highest one month ago at midnight?

9) Does the constellation that was highest in the sky at midnight a month ago now rise earlier or later than it rose last month? Explain your reasoning.

Part II: Daily Differences

Figure 3 shows the same Earth–Sun view as before and the bright star Betelgeuse, which is between Taurus and Gemini.

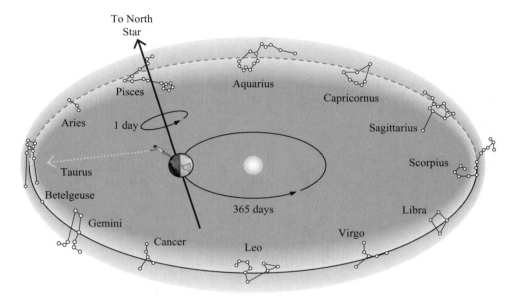

Figure 3

10) Imagine last night you saw the star Betelgeuse just starting to rise on your eastern horizon at 7:15 P.M. At 7:15 P.M. tonight, will Betelgeuse be above, below, or exactly on your eastern horizon?

11) Two students are discussing their answers to Question 10.

   Student 1: *Earth makes one complete rotation about its axis each day so Betelgeuse will rise at the same time every night. It will therefore be exactly on the eastern horizon.*

   Student 2: *No. Because Earth goes around the Sun, the constellation Taurus rises earlier each month and so it must rise a little bit earlier each night, too. Betelgeuse must do the same thing. Tonight it would rise a little before 7:15 and be above the eastern horizon by 7:15.*

   Do you agree or disagree with either or both of the students? Explain your reasoning.

Part I: Solar Day

Figure 1 shows a top-down view of the Earth–Sun system. Arrows indicate the directions of the rotational and orbital motions of Earth. For the observer shown, the Sun is highest in the sky at noon.

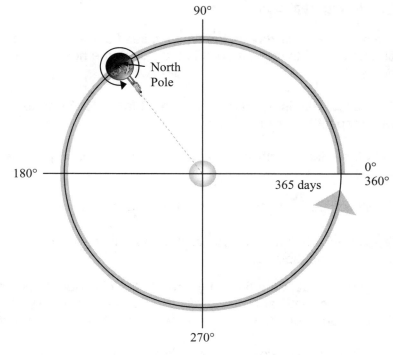

Figure 1

1) Earth orbits the Sun in a counterclockwise direction once every 365 days. Approximately how many degrees does Earth move along its orbit in one day?

2) As Earth orbits the Sun, it also rotates in a counterclockwise direction about its axis as shown in Figure 1. We define 24 hours as the time from when the Sun is highest in the sky one day to when it is highest in the sky the next day. How many degrees does Earth rotate about its axis in exactly 24 hours: 360°, slightly less than 360°, or slightly more than 360°?

3) How long does it take Earth to rotate exactly 360°: slightly less than 24 hours, 24 hours, or slightly more than 24 hours?

4) Two students are discussing their answers to Questions 2 and 3.

   Student 1: *Earth rotates about its axis once every 24 hours, and one rotation equals 360°.*

   Student 2: *No. When Earth has gone around 360° it has also moved a small amount counterclockwise around the Sun, which means the Sun is not yet at its highest point. Earth must spin a little bit more for the Sun to reach its highest point.*

   Do you agree or disagree with either or both of the students? Explain your reasoning.

Part II: Sidereal Day

We define a **solar day** as the time it takes for the Sun to go from its highest point in the sky on one day to its highest point in the sky on the next day, and we divide that time into 24 hours.

A **sidereal day** is defined as the time it takes for Earth to rotate *exactly* 360° about its axis with respect to the distant stars.

5) When does Earth rotate a greater amount, during a solar day or during a sidereal day?

6) Which takes a shorter amount of time, a solar day or a sidereal day?

Note: Since Earth rotates more than 360° in a solar day, a sidereal day is about 4 minutes shorter than a solar day.

Imagine that at some time in the future the direction that Earth orbits the Sun is somehow reversed so that Earth now orbits the Sun approximately 1° *clockwise* each day. However, the rotation about its own axis remains counterclockwise at the same rate.

7) In the space below, create a sketch similar to Figure 1 to depict this imaginary situation.

8) Through how many degrees will Earth now rotate in a *sidereal* day?

9) Through how many degrees will Earth now rotate in a *solar* day?

10) Which is now longer, the solar or the sidereal day?

11) Is a sidereal day now longer, shorter, or the same length as a sidereal day was before we changed Earth's orbital direction?

12) Is a solar day now longer, shorter, or the same length as a solar day was before we changed Earth's orbital direction?

For all parts of this activity, it is helpful to imagine that the stars are so bright (or our Sun so dim) that the stars can be seen during the day so that your sky might appear as in Figure 1.

Part I: Daily Motion

On December 1, at noon, you are looking toward the south and see the Sun among the stars of the constellation Scorpius as shown in Figure 1.

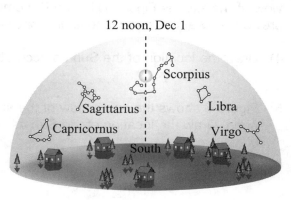

Figure 1

1) At 3 P.M. that afternoon, will the Sun appear among the stars of the constellation Capricornus, Sagittarius, Scorpius, Libra, or Virgo?

2) Two students are discussing their answers to Question 1.

   Student 1: *The Sun moves from the east through the southern part of the sky and then to the west. By 3 P.M. it will have moved from being high in the southern sky to the west into the constellation Libra.*

   Student 2: *You're forgetting that stars and constellations, like those in Figure 1, will rise in the east, move through the southern sky, and then set in the west just like the Sun. So the Sun will still be in Scorpius at 3 P.M.*

   Do you agree or disagree with either or both of the students? Explain your reasoning.

Recall that in the celestial sphere model, the stars' daily motions result from the rotation of the celestial sphere.

3) Is it reasonable to account for the Sun's **daily motion** by assuming that the Sun is at a fixed position on the celestial sphere (in this case in the location of the constellation Scorpius) and is carried along its path across the sky by the sphere's rotation? Explain why or why not.

Part II: Monthly Changes

By careful observation of the Sun's position in the sky throughout the year, we find that the celestial sphere rotates slightly more than 360° every 24 hours. Figure 2 shows the same view of the sky (as Figure 1) but on December 2 at noon. For comparison, the view from the previous day at the same time is also shown in gray.

4) Draw the location of the Sun as accurately as possible in Figure 2.

5) Figure 3 shows the same view of the sky (as Figure 1) one month later on January 1 at noon. Draw the location of the Sun as accurately as possible in this figure.

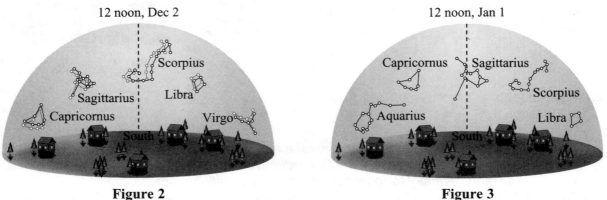

Figure 2   **Figure 3**

6) Two students are discussing their answers to Questions 4 and 5.

> **Student 1:** *The Sun will always lie along the dotted line in the figures when it's noon.*
>
> **Student 2:** *But, we saw in Question 3 that the Sun's motion can be modeled by assuming it is stuck to the celestial sphere. The Sun must, therefore, stay in Scorpius.*
>
> **Student 1:** *If that were true, then by March the Sun would be setting at noon. The Sun must shift a little along the celestial sphere each day so that in 30 days it has moved toward the east into the next constellation.*

Do you agree or disagree with either or both of the students? Explain your reasoning.

7) Why is it reasonable to think of the Sun as attached to the celestial sphere over the course of a single day, as suggested in Question 3, even though we know from Questions 5 and 6 that the Sun's position is not truly fixed on the celestial sphere?

Part III: The Ecliptic

The zodiacal constellations were of special interest to ancient astronomers because these are the constellations through which the Sun moves throughout the year. This was incorporated into their celestial sphere model by having the Sun loosely fixed to the celestial sphere but allowing it to slip a small amount each day. The Sun's position on the celestial sphere (among the stars in the constellation Scorpius) on December 1 is shown in Figure 4.

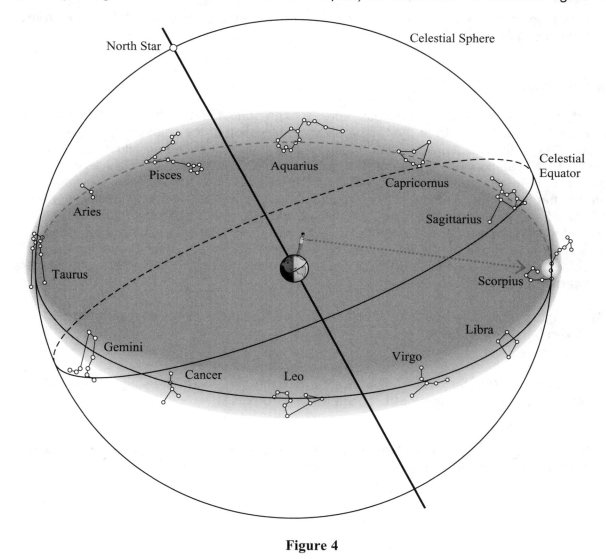

Figure 4

8) On Figure 4, draw where the Sun will be located on the celestial sphere on January 1. Label this position "Jan 1."

9) On Figure 4, for the other constellations, draw in the Sun and label the constellation with the approximate date that the Sun will be located there.

The line drawn through these constellations, tracing out the Sun's annual path, is called the **ecliptic.**

10) Label the ecliptic in Figure 4.

11) About how many times does the celestial sphere rotate in the time it takes the Sun to move between two adjacent constellations (i.e., 1/12 of the way around) along the ecliptic?

12) How long does it take the Sun to make one complete trip around the ecliptic (i.e., from Scorpius to Scorpius)?

Part IV: Wrap-Up

It is important to realize that the ecliptic represents an *annual* drift of the Sun and does not represent the daily path of the Sun. Instead, the rotation of the celestial sphere is responsible for the Sun's daily motion through the sky. Also, since the ecliptic is tilted with respect to the rotation axis of the celestial sphere, the ecliptic slowly "wobbles" as the celestial sphere rotates. The Sun's position on the ecliptic is only important in deciding whether the Sun's daily path will carry it high in the sky (summer) or low in the sky (winter). In Figure 5a, the Sun's position along the ecliptic and its path for one day (dashed line) are shown for two different dates: December 1 (in Scorpius) and June 1 (in Taurus). Figures 5b, 5c, and 5d show the path of the Sun and the wobble of the ecliptic at six-hour intervals as the celestial sphere rotates. Study these figures, carefully noting that the ecliptic and Sun are both carried by the celestial sphere.

13) On Figure 5d, label the ecliptic (Sun's annual path) and the Sun's daily path for December 1 and June 1.

14) Which Figure (5a, 5b, 5c, or 5d) shows the Sun at noon, low in the southern sky, when it would be among the stars of the constellation Scorpius?

15) Which Figure (5a, 5b, 5c, or 5d) shows the Sun at noon, high in the southern sky, when it would be among the stars of the constellation Taurus?

December 1 (Sun in Scorpius)
and
June 1 (Sun in Taurus)

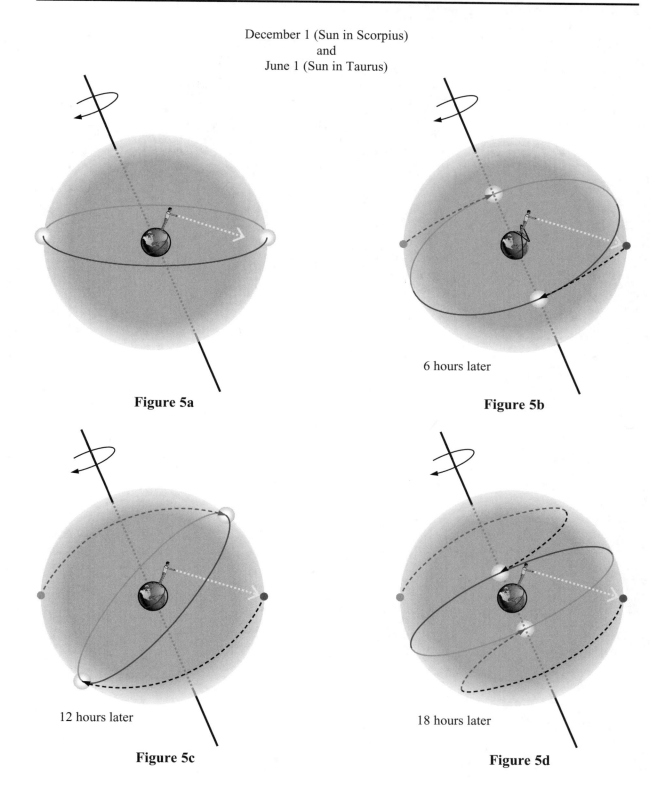

Figure 5a

6 hours later

Figure 5b

12 hours later

Figure 5c

18 hours later

Figure 5d

Consider the overhead-view star map for July at midnight shown in Figure 1. In particular, notice that the directions of north and east have been identified and that the names of different star groups (constellations) have been provided.

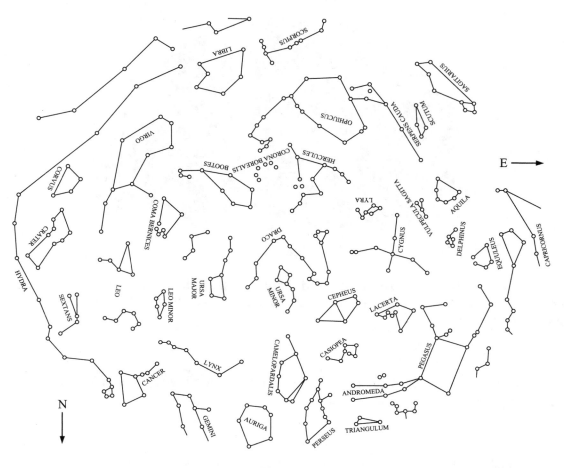

Figure 1

1) Which star group will appear highest in the night sky at this particular time?

2) Figure 2 shows a south-facing horizon view star map for July at midnight. What is the name of the star group that appears highest in the sky on this south-facing horizon view star map? (Hint: refer to the names provided in Figure 1.)

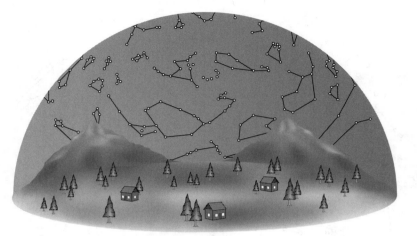

Figure 2

3) How would you have to hold, rotate, fold, and/or change the overhead-view star map shown in Figure 1 so that it could be used as a south-facing star map like the one provided in Figure 2?

4) How would your answer to the previous question change if you wanted to use the star map from Figure 1 as a north-facing map?

5) Do you still agree with your answer to Question 1? Why or why not?

6) When looking at the overhead-view star map from Figure 1,

   a) on what part of the map (left, right, top, bottom, or center) is the star group that will appear highest in the night sky? What is the name of this star group?

   b) on what part of the map (left, right, top, bottom, or center) is the star group that will appear near the southern horizon? What is the name of this star group?

   c) on what part of the map (left, right, top, bottom, or center) is the star group that will appear near the eastern horizon? What is the name of this star group?

Part I: Equal Area in Equal Time Intervals

Kepler's second law of planetary motion states that a line joining a planet and the Sun sweeps out equal amounts of area in equal intervals of time.

Imagine the situation shown at the right in which a planet is moving in a perfectly circular orbit around its companion star. Note that the time between each position shown is exactly one month.

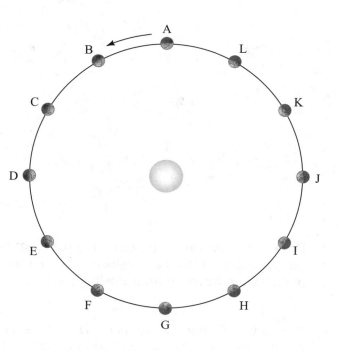

1) Does this planet obey Kepler's second law? How do you know?

2) If you were carefully watching this planet during the entire orbit, would the speed of the planet be increasing, decreasing, or staying the same? How do you know?

In the drawing below, a planet that obeys Kepler's second law is shown at nine different locations (A–I) during the planet's orbit around its companion star.

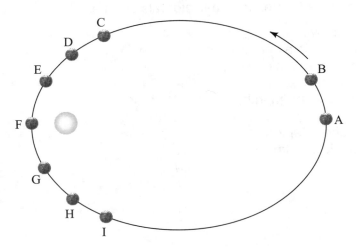

3) Draw two lines: one connecting the planet at Position A to the star and a second line connecting the planet at Position B to the star. Shade in the area swept out by the planet when traveling from Positions A to B.

4) Pick any two planet positions (C, D, E, F, G, H, I)—note, they do not have to be consecutive—that you could use to construct a swept-out area that would have approximately the same area as the one you shaded in for Question 3. Shade in the second swept-out area using the planet positions that you chose. Note: Your shaded area needs to be only roughly the same size; no calculations or quantitative estimates are required.

5) How would the time it takes the planet to travel from Position A to Position B compare (greater than, less than, or equal to) to the time it takes to travel between the two positions you selected in Question 4? Explain your reasoning.

6) During which of the two time intervals for which you sketched the shaded areas in Questions 3 and 4 is the distance traveled by the planet greater?

7) During which of the two time intervals for which you sketched the shaded areas in Questions 3 and 4 would the planet be traveling faster? Explain your reasoning.

Part II: Kepler's Second Law and the Speed of the Planets

The drawing below shows another planet's orbit. In this case, the twelve positions shown (A–L) are each exactly one month apart. As before, the planet shown obeys Kepler's second law.

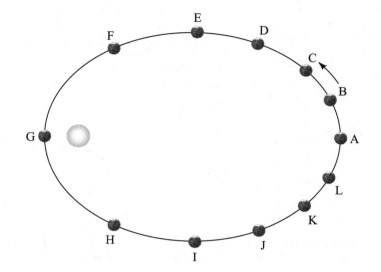

8) Does the planet appear to be traveling the same distance each month?

9) At which position would the planet have been traveling the fastest? The slowest? Explain your reasoning.

10) At Position D, is the speed of the planet increasing or decreasing as time goes on? Explain your reasoning.

11) Provide a concise statement that describes the relationship that exists between a planet's orbital speed and the planet's distance from its companion star.

Part III: Kepler's Second Law and Eccentricity

Consider the table below listing the orbit eccentricities for objects in the solar system. Recall that an orbit with an eccentricity of zero is perfectly circular whereas the highly elliptical orbits shown in Parts I and II would have a high eccentricity of approximately 0.90.

Object	Eccentricity of Orbit
Mercury	0.206
Venus	0.007
Earth	0.016
Mars	0.093
Jupiter	0.048
Saturn	0.054
Uranus	0.047
Neptune	0.008
Pluto	0.248

12) Which of the three orbits shown below (A, B, or C) would you say most closely matches the shape of Earth's orbit around the Sun? Explain your reasoning.

13) Which of the listed objects would experience the largest change in orbital speed and which would experience the smallest change in orbital speed?

14) Describe the extent to which you think Earth's orbital speed changes throughout a year? Explain your reasoning.

Kepler's third law describes the relationship between how long it takes a planet to orbit a star (orbital period) and how far away that planet is from the star (orbital distance). In this activity, we investigate an imaginary planetary system that has an average star, like the Sun, at the center and two planets. A huge Jupiter-like, Jovian planet named Esus orbits close to the star, while a small Earth-like, terrestrial planet named Sulis is in an orbit far away around the star. Use this information when answering the next four questions. If you're not sure of the correct answers to Questions 1–4, just take a guess. We'll return to these questions later in this activity.

1) Which of the two planets (Esus or Sulis) do you think will move around the central star in the least amount of time? Explain your reasoning.

2) If Esus and Sulis were to switch positions, would your answer to Question 1 change? If so, how? If not, why not?

3) Do you think the orbital period for Esus would increase, decrease, or stay the same if its mass were increased? Explain your reasoning.

4) Imagine both Esus and Sulis were in orbit around the same central star at the same distance and that their orbital positions would never intersect (so that they would never collide). Which of the two planets (Esus or Sulis) do you think will move around the central star in the least amount of time? Explain your reasoning.

The graph below illustrates how the orbital period (expressed in years) and orbital distance (expressed in astronomical units, AU) of a planet are related.

5) According to the graph, would you say that the orbital period of planets appears to increase, decrease, or stay the same as their orbital distance is increased?

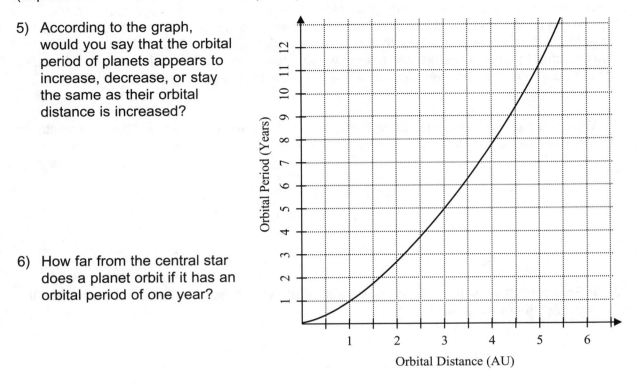

6) How far from the central star does a planet orbit if it has an orbital period of one year?

7) How long does it take a planet to complete one orbit if it is twice the distance from the central star as the planet described in Question 6?

8) Based on your results from Questions 6 and 7, which of the following best describes how a planet's orbital period will change (if at all) when its distance to the central star is doubled? Circle your choice.

   a) The planet's orbital period will decrease by half.

   b) The planet's orbital period will not change.

   c) The planet's orbital period will double.

   d) The planet's orbital period will more than double.

In the table below we have provided the orbital distances, orbital periods, and masses for the six planets closest to the Sun.

Planet	Orbital Distance (in astronomical units—AU)	Orbital Period (in years)	Planet Mass (in units of Earth's mass)
Mercury	0.38	0.24	0.06
Venus	0.72	0.61	0.82
Earth	1.0	1.0	1.0
Mars	1.52	1.88	0.11
Jupiter	5.20	11.86	318
Saturn	9.54	29.46	95.2

9) What is the name of the planet that you identified the orbital distance for in Question 6?

10) Using the information provided in the table above and on the graph on the previous page, which of the answers below best describes how a planet's mass will affect its orbital period. Circle your choice.

a) Planets that have small masses have longer orbital periods than planets with large masses.

b) Planets with the same mass will also have the same orbital period.

c) Planets that have large masses have longer orbital periods than planets with small masses.

d) A planet's mass does not affect the orbital period of a planet.

Explain your reasoning and cite a specific example from the table or graph to support your choice.

11) A student in your class makes the following comment about the relationship between the location of planets in our solar system and their orbital period and mass.

Student: *As we look at planets farther away from the Sun than Mercury we see that their distances get bigger and that the mass of the planets is also getting larger. So I think that the farther away a planet is from the Sun, the more massive it will be and the longer it will take to go around the Sun.*

Which planet listed in the table above illustrates the fact that the student's reasoning is incorrect? Explain your reasoning.

12) Review your answers to Questions 1–4. Do you still agree with the answers you provided? If not, describe (next to your original answers) how you would change the answers you gave initially.

Part I: The Force of Gravity

Newton's law of universal gravitation describes the attractive gravitational force that exists between any two bodies with the following equation:

$$F_G = \frac{GMm}{r^2}$$

G is the gravitational constant (which for this activity you can assign a value of 1). **M** and **m** are the masses of the two objects attracting one another, and **r** is the distance from the center of one object to the center of the other object.

1) Given that Earth is much larger and more massive than the Moon, how does the strength of the gravitational force that the Moon exerts on Earth compare to the gravitational force that Earth exerts on the Moon? Explain your reasoning.

2) Consider the following debate between two students about their answer to the previous question.

Student 1: *I thought that whenever one object exerts a force on a second object, the second object also exerts a force that is equal in strength, but in the other direction. So even though Earth is bigger and more massive than the Moon, they still pull on each other with a gravitational force of the same strength, just in different directions.*

Student 2: *I disagree. I said that Earth exerts the stronger force because it is way bigger than the Moon. Because its mass is bigger, the gravitational force Earth exerts has to be bigger too. I think you are confusing Newton's third law with the law of gravity.*

Do you agree or disagree with either or both of the students? Explain your reasoning.

3) How would the strength of the force between the Moon and Earth change if the mass of the Moon were somehow made two times greater than its actual mass?

Part II: Force–Distance Relationship

In the picture below, a spaceprobe traveling from Earth to Mars is shown at the halfway point between the two (not to scale).

🔴 Mars

🛰

 Earth

4) On the diagram, clearly label the location where the spaceprobe would be when the gravitational force by Earth on the spaceprobe is strongest? Explain your reasoning.

5) On the diagram, clearly label the location where the spaceprobe would be when the gravitational force by Mars on the spaceprobe is strongest. Explain your reasoning.

6) Where would the spaceprobe experience the strongest net (or total) gravitational force exerted on it by Earth and Mars? Explain your reasoning.

7) When the spacecraft is at the halfway point, how does the strength of the gravitational force on the spaceprobe by Earth compare with the strength of the gravitational force on the spaceprobe by Mars? Explain your reasoning.

8) Two students are discussing their answer to the previous question.

Student 1:   *Since the spaceprobe is exactly halfway between Earth and Mars, the strength of the gravitational forces would be the same since the distances are the same.*

Student 2:   *You're right that the distances are the same, but you're forgetting about mass. The combined mass of the spaceprobe and Earth is much bigger than the combined mass of the spaceprobe and Mars. So, since the distances are the same, the strength of the gravitational force on the spaceprobe by Earth has to be bigger than the strength of the gravitational force on the spaceprobe by Mars.*

Do you agree or disagree with either or bother of the students? Explain your reasoning.

9) If the spaceprobe had lost all ability to control its motion and was sitting at rest at the midpoint between Earth and Mars, would the spacecraft stay at the midpoint or would it start to move?

If you think it stays at the midpoint, explain why it would not move.

If you think it would move, then: (a) Describe the direction it would move; (b) describe if it would speed up or slow down; (c) describe how the net (or total) force on the spaceprobe would change during this motion; and (d) identify when/where the spaceprobe would experience the greatest acceleration.

10) Imagine that you need to completely stop the motion of the spaceprobe and have it remain at rest while you perform a shutdown and restart procedure. You have decided that the best place to carry out this procedure would be at the position where the net (or total) gravitational force on the spaceprobe by Mars and Earth would be zero. On the diagram, label the location where you would perform this procedure. (Make your best guess; there is no need to perform any calculations here.) Explain the reasoning behind your choice.

11) Your weight on Earth is simply the gravitational force between you and Earth. Would your weight be more, less, or the same on Mars? Explain your reasoning.

1) Which value, apparent magnitude, or absolute magnitude, do you think:

   a) tells us how bright an object will appear from Earth?

   b) tells us about the object's actual brightness?

2) The full Moon has an apparent magnitude of −12.6, and when Mars is at its brightest in the night sky, its apparent magnitude is +2.0.

   a) Which of the two objects has the bigger apparent magnitude number?

   b) Which object will look brighter from Earth, the full Moon or Mars? How do you know?

   c) If a new object were discovered that looked even dimmer from Earth than Mars does, make up a possible apparent magnitude number for it.

3) Consider the following debate between two students.

   Student 1: *I think a star with an apparent magnitude number of −2.0 would look brighter than a star with an apparent magnitude number of +1.0.*

   Student 2: *I disagree. You don't understand the number scale for apparent and absolute magnitude. The bigger the number the brighter the star. So the +1.0 star would look brighter than the −2.0 star.*

   Do you agree or disagree with either or both of the students? Explain your reasoning.

4) Star Y appears much brighter than Star Z when viewed from Earth, but is found to actually give off much less light. Assign a set of possible values for the apparent and absolute magnitudes of these stars that would be consistent with the information given in the previous statement. Explain your reasoning.

5) The star Lee has an apparent magnitude of 0.1 and is located about 250 parsecs away from Earth. Which of the following is most likely the absolute magnitude for Lee?

   a) −6.9

   b) 0.1

   c) 7.1

   Explain your reasoning.

6) Refer to the following table for Questions 6a–6d:

	Apparent Magnitude	Absolute Magnitude
Star A:	1	1
Star B:	1	2
Star C:	5	4
Star D:	4	4

   a) Which object appears brighter from Earth: Star C or Star D? Explain your reasoning.

   b) Which object is actually brighter: Star A or Star D? Explain your reasoning.

c) For Stars A–D, state whether the star is closer than, farther than, or exactly 10 parsecs away from Earth. Explain your reasoning.

d) Would the apparent magnitude number of Star A increase, decrease, or stay the same, if it were located at a distance of 40 parsecs? What about the absolute magnitude number? Explain your reasoning.

7) Star F is known to have an apparent magnitude of −26.7 and an absolute magnitude of 4.8. Where might this star be located? What is the name of this star? Explain your reasoning.

Part I: Stars in the Sky

Consider the diagram to the right.

1) Imagine that you are looking at the stars from Earth in January. Use a straightedge or a ruler to draw a straight line from Earth in January, through the Nearby Star (Star A), out to the Distant Stars. Which of the distant stars would appear closest to Star A in your night sky in January? Circle this distant star and label it "Jan."

2) Repeat Question 1 for July and label the distant star "July."

3) In the box below, the same distant stars are shown as you would see them in the night sky. Draw a small × to indicate the position of Star A as seen in January and label it "Star A Jan."

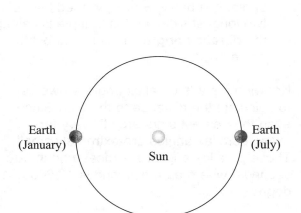

4) In the same box, draw another × to indicate the position of Star A as seen in July and label it "Star A July."

5) Describe how Star A would appear to move among the distant stars as Earth orbits the Sun counterclockwise from January of one year, through July, to January of the following year.

The apparent motion of nearby objects relative to distant objects, which you just described, is called **parallax.**

6) Consider two stars (C and D) that both exhibit parallax. If Star C appears to move back and forth by a greater amount than Star D, which star do you think is actually closer to you? If you're not sure, just take a guess. We'll return to this question later in this activity.

Part II: What's a Parsec?

Consider the diagram to the right.

7) Starting from Earth in January, draw a line through Star A to the top of the page.

8) There is now a narrow triangle, created by the line you drew, the dotted line provided in the diagram, and the line connecting Earth and the Sun. The small angle, just below Star A, formed by the two longest sides of this triangle is called the **parallax angle** for Star A. Label this angle "p_A."

Knowing a star's parallax angle allows us to calculate the distance to the star. Since even the nearest stars are still very far away, parallax angles are extremely small. These parallax angles are measured in "arcseconds" where an arcsecond is 1/3600 of 1 degree.

To describe the distances to stars, astronomers use a unit of length called the **parsec.** One parsec is defined as the distance to a star that has a **par**allax angle of exactly 1 arc**sec**ond. The distance from the Sun to a star 1 parsec away is 206,265 times the Earth–Sun distance or 206,265 AU. (Note that the diagram to the right is not drawn to scale.)

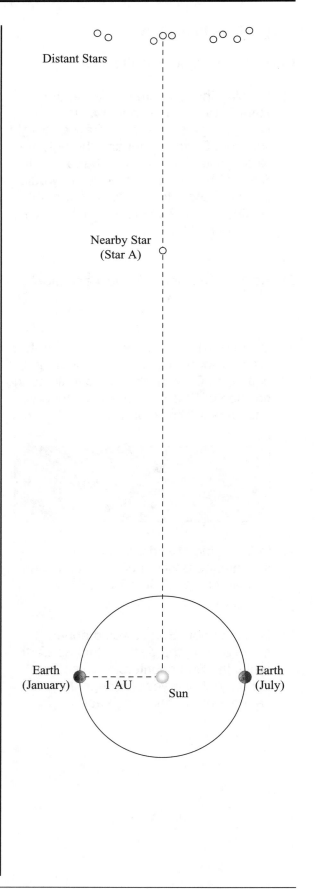

Distant Stars

Nearby Star (Star A)

Earth (January)   1 AU   Sun   Earth (July)

9) If the parallax angle for Star A (p_A) is 1 arcsecond, what is the distance from the Sun to Star A? (Hint: Use parsec as your unit of distance.) Label this distance on the diagram.

10) Is a parsec a unit of length or a unit of angle? It can't be both.

Note: Since the distance from the Sun to even the closest star is so much greater than 1 AU, we can consider the distance from Earth to a star and the distance from the Sun to that star to be approximately equal.

Part III: Distances

11) Consider the following debate between two students regarding the relationship between parallax angle and the distance we measure to a star.

Student 1: *If the distance to the star is more than 1 parsec, then the parallax angle must be more than 1 arcsecond. So a star that is many parsecs away will have a large parallax angle.*

Student 2: *If we drew a diagram for a star that was much more than 1 parsec away from us, the triangle in the diagram would be pointier than the one we just drew in Part II. That should make the parallax angle smaller for a star farther away.*

Do you agree or disagree with either or both of the students? Explain your reasoning.

12) On your diagram from Part II, draw a second star along the dotted line farther from the Sun than Star A and label this faraway star "Star B." Repeat steps 7 and 8 from Part II, except label the parallax angle for this Star B with p_B.

13) Which star, the closer one (Star A) or the farther one (Star B), has the larger parallax angle?

14) Check your answers to Questions 6 and 11 and resolve any discrepancies.

Part I: Angular Measurement

Imagine that you are standing in an open field. While facing south, you see a house in the distance. If you turn your head and look directly east (to your left), you see a barn in the distance.

1) What is the angle between you, the house, and the barn? (Hint: If you point at the barn with one arm and point at the house with your other arm, what angle do your arms make?)

2) You see the Moon on the horizon just above the barn in the east and also see a bright star directly overhead. What is the angle between you, the Moon, and the overhead star?

3) Compare your answers for the barn–house angle from Question 1 and the Moon–star angle from Question 2. Are they the same?

4) Do the angles from above tell you anything about the actual distance between the barn and house or the Moon and star?

We are often unable to **directly** measure distances to faraway objects in our night sky. However, we can obtain the distances to relatively nearby stars by using their parallax angles. Because even these stars are very far away (up to about 500 parsecs), the parallax angles for these stars are very small. They are measured in units of **arcseconds**, where 1 arcsecond is 1/3600 of 1 degree. To give you a sense of how small this angle is, the thin edge of a credit card, when viewed from one football field away, covers an angle of about 1 arcsecond.

Part II: Finding Stellar Distance Using Parallax

Consider the starfield drawing shown in Figure 1. This represents a tiny patch of our night sky. In this drawing we will imagine that the angle separating Stars A and B is just 1/2 of an arcsecond.

Figure 1

In Figure 2 (see the final page of the activity) there are drawings of this starfield taken at different times during the year. One star in the field moves back and forth across the star field (exhibits parallax) with respect to the other, more distant stars.

5) Using Figure 2, determine which star exhibits parallax. Circle that star on each picture in Figure 2.

6) In Figure 1, draw a line that shows the range of motion for the star you saw exhibiting parallax in the drawings from Figure 2. Label the end points of this line with the months when the star appears at those end points.

7) How many times bigger is the separation between Stars A and B compared to the distance between the end points of the line showing the range of the motion for the star exhibiting parallax?

8) Recall that Stars A and B have an angular separation of 1/2 of an arcsecond in Figure 1. Consider two more stars (C and D) that are separated **twice** as much as Stars A and B. What is the angular separation between Stars C and D in arcseconds?

9) What is the angular separation between the end points that you marked in Figure 1 for the nearby star exhibiting parallax?

Note: We define a star's **parallax angle as half** the angular separation between the end points of the star's angular motion.

10) What is the parallax angle for the nearby star exhibiting parallax from Question 9?

Note: We define 1 **parsec** as the distance to an object that has a **par**allax angle of 1 arc**sec**ond. For a star with a parallax angle of 2 arcseconds, the distance to the star from Earth would be 1/2 of a parsec.

11) For a star with a parallax angle of 1/2 of an arcsecond, what is its distance from us?

12) For a star with a parallax angle of 1/4 of an arcsecond, what is its distance from us?

13) What is the distance from us to the nearby star exhibiting parallax in the drawings from Figure 2? (Hint: Consider your answer to Question 10.)

   a) 1 parsec

   b) 2 parsecs

   c) 4 parsecs

   d) 8 parsecs

   e) 16 parsecs

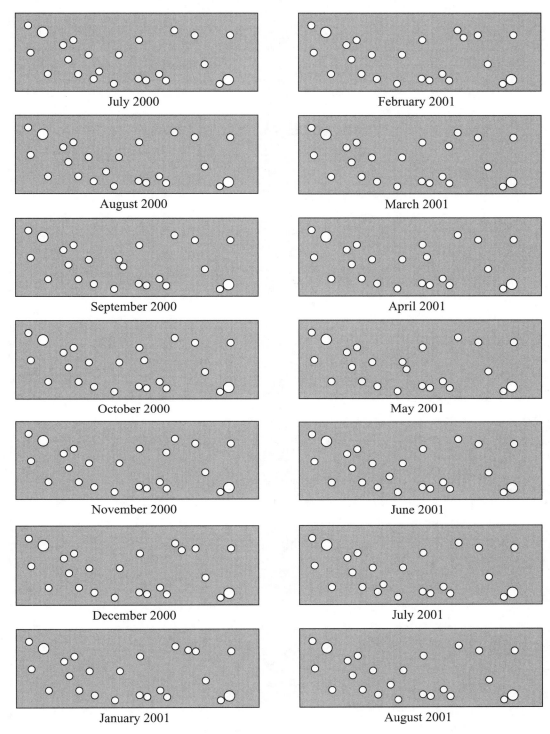

Figure 2

Part I: Magnitudes and Star Distances

Below is a table of four stars along with their apparent and absolute magnitudes. Use this table to answer the following questions.

	Apparent Magnitude	Absolute Magnitude	Distance
Star A:	0	0	
Star B:	0	2	
Star C:	5	4	
Star D:	4	4	

1) Which object appears brighter from Earth: Star C, Star D, or neither? Explain your reasoning.

2) Which object is more luminous: Star C, Star D, or neither? Explain your reasoning.

3) Star B has an apparent magnitude of 0, which tells us how bright it appears from Earth at its true location. Star B has an absolute magnitude of 2, which tells us how bright it would appear if it were at a distance of 10 parsecs (about 33 light-years).

   Where would Star B appear brighter, in its true location or if it were at a distance of 10 parsecs? Explain your reasoning.

4) Is Star B *closer than 10 parsecs, farther than 10 parsecs,* or *exactly 10 parsecs* away? Record your answer in the table above, and explain your reasoning.

5) Complete the remaining blanks in the distance column of the above table and check your answers with another group. Explain your reasoning for the distances you recorded for Stars A, C, and D.

Part II: Spectroscopic Parallax

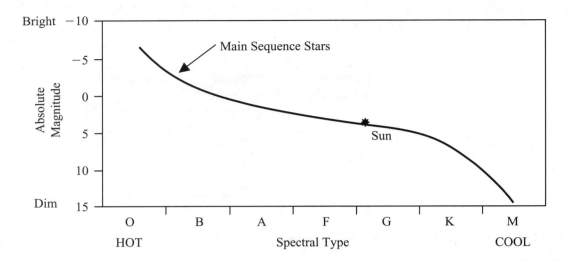

Below is a table giving both the apparent magnitude and spectral type for five **main sequence** stars. For each star, do the following:

6) Using the above H–R diagram, estimate the absolute magnitude for each star and write your answer in the absolute magnitude column of the table below.

7) Complete the distance column in the table below by classifying each star as being **closer, slightly farther,** or much **farther** than 10 parsecs away. This procedure, called spectroscopic parallax, provides astronomers with another way to measure the distance to stars.

Star	Apparent Magnitude	Spectral Type	Absolute Magnitude	Distance Estimate
Rigel Kentaurus	0.0	G2		
Vega	0.04	A0		
Rigel B	6.6	B9		
Achernar	0.5	B3		
Tau Scorpius	2.8	B0		

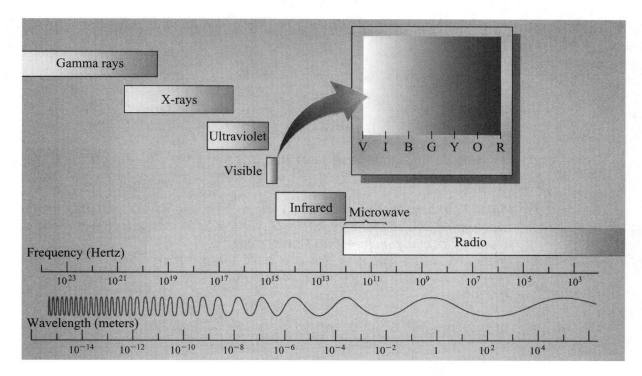

1) The electromagnetic spectrum of light is often arranged in terms of frequency. Which one of the following has the highest frequency (*circle one*)?

visible light	microwaves	infrared light	gamma rays
radio waves	X-rays	ultraviolet light	

2) The electromagnetic spectrum of light can also be arranged in terms of wavelengths. Which one of the following has the longest wavelength (*circle one*)?

visible light	X-rays	ultraviolet light	infrared light
gamma rays	microwaves	radio waves	

3) Which of the following types of light travels at the fastest speed (*circle your answer(s)*)? Explain your answer:

ultraviolet light	X-rays	gamma rays	visible light
microwaves	radio waves	infrared light	

4) Another property of light is the energy. Which of the following has the greatest energy (*circle one*)?

ultraviolet   X-rays   gamma   visible
   light   rays   light
          microwaves   radio   infrared
                                    waves   light

5) Consider the following discussion between two students about the different properties of light.

   Student 1:   *I think I get how light works. If you look at the chart of the electromagnetic spectrum, it shows that light with a higher frequency will also have a long wavelength. But it all has the same speed.*

   Student 2:   *I disagree. If one type of light has a lot of energy and a high frequency, it will have a faster speed than light that has a lower energy and a low frequency.*

   Do you agree or disagree with either or both of the students? Explain your reasoning.

6) Complete the following sentence describing the relationship between the energy, frequency, and wavelength of light, using the words *highest, lowest, longest,* and/or *shortest.*

   The portion of the electromagnetic spectrum of light with the **greatest** *energy has the* _____ *frequency and the* _____ *wavelengths.*

7) The visible light portion of the electromagnetic spectrum of light is often subdivided into the colors of red, orange, yellow, green, blue, indigo, and violet (*sometimes referred to as ROY G BIV*). Using the words *greatest, least, highest, lowest, fastest, slowest, longest,* and *shortest*, write a sentence or two that describes how light at the red end of the visible portion of the spectrum and light at the violet end of the visible light portion of the spectrum compare in terms of their energy, frequency, speed, and wavelength.

8) For each statement (a–d) provided below, circle the word choice that correctly describes how the two forms of light compare.

   a) Infrared light has <u>greater / less</u> energy than ultraviolet light.

   b) X-ray photons have <u>longer / shorter</u> wavelengths than gamma ray photons.

   c) Visible electromagnetic radiation has a <u>higher / lower</u> frequency than radio wave electromagnetic radiation.

   d) Infrared light has a <u>faster / slower / same</u> speed than microwave light.

9) Of all the types of light the Sun gives off, it emits the greatest amount of light at visible light wavelengths. If the Sun were to cool off dramatically and as a result start giving off mainly light at wavelengths longer than visible light, how would the frequency, energy, and speed of this light given off by the Sun also be different? Explain your reasoning.

LECTURE-TUTORIALS FOR INTRODUCTORY ASTRONOMY
                                                                    THIRD EDITION

The drawing below illustrates the amount that different wavelengths of light are able to penetrate down through Earth's atmosphere. The shaded regions are used in this drawing to depict different layers in Earth's atmosphere. Notice that the atmosphere can be completely transparent to light at some wavelengths (all three lines passing through the atmosphere to the surface of Earth) and yet can also completely absorb other wavelengths of light (all three lines stopping in the atmosphere before reaching Earth's surface).

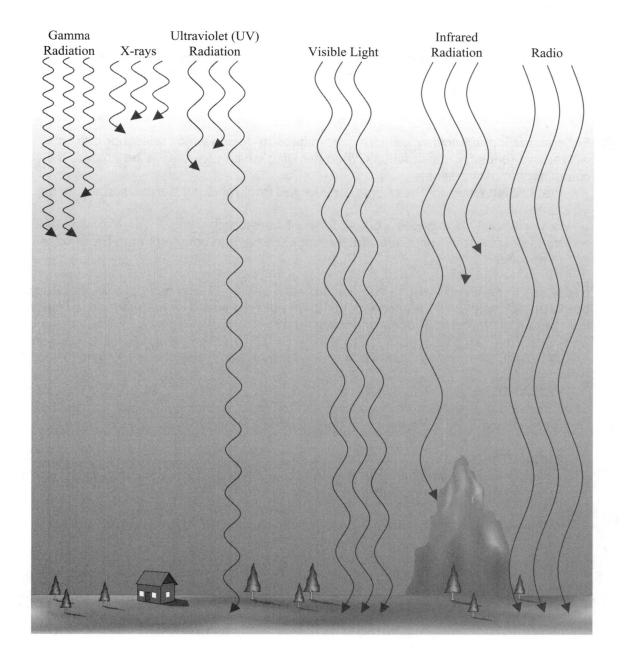

1) Which, if any, of the different wavelengths of light (electromagnetic radiation) shown in the image on the previous page are able to *completely* penetrate Earth's atmosphere and reach the surface?

2) Which, if any, of the different wavelengths of light (electromagnetic radiation) shown in the image on the previous page only *partially* penetrate Earth's atmosphere and reach the surface?

3) Which, if any, of the different wavelengths of light (electromagnetic radiation) shown in the image on the previous page are *completely* absorbed in Earth's atmosphere and never reach the surface?

4) Federal funding agencies must form committees to decide which telescope projects will receive funds for construction. When deciding which projects will be funded, the committees must consider:
   • that certain wavelengths of light are blocked from reaching Earth's surface by the atmosphere,
   • how efficiently telescopes work at different wavelengths, and
   • that telescopes in space are much more expensive to construct than Earth-based telescopes.

   Use these three criteria when you consider each pairing of telescope proposals listed below (a–d). State which proposal out of each pair you would choose to fund. Explain the reasoning behind your decision *for each pair*.

   a) Which of the two proposals described below would you choose to fund?

   Project Delta:
   A gamma ray wavelength telescope, located in Antarctica, which will be used to look for evidence to indicate the presence of a black hole.

   Project Theta:
   A visible wavelength telescope, located on a university campus, which will be used in the search for planets outside the solar system.

   Explain your reasoning.

b) Which of the two proposals described below would you choose to fund?

Project Beta:
An X-ray wavelength telescope, located near the North Pole, which will be used to examine the Sun.

Project Alpha:
An infrared wavelength telescope, placed on a satellite in orbit around Earth, which will be used to view supernovae.

Explain your reasoning.

c) Which of the two proposals described below would you choose to fund?

Project Rho:
A UV wavelength telescope, placed high atop Mauna Kea in Hawaii at 14,000 ft above sea level, which will be used to look at distant galaxies.

Project Sigma:
A visible wavelength telescope, placed on a satellite in orbit around Earth, which will be used to observe a pair of binary stars located in the constellation Ursa Major.

Explain your reasoning.

d) Which of the two proposals described below would you choose to fund?

Project Zeta:
A radio wavelength telescope, placed on the floor of the Mojave Desert, which will be used to detect potential communications from distant civilizations outside our solar system.

Project Epsilon:
An infrared wavelength telescope, located in the high-elevation mountains of Chile, which will be used to view newly forming stars (protostars) in the Orion nebula.

Explain your reasoning.

Part I: Luminosity, Temperature, and Size

Imagine you are comparing the ability of electric hot plates of different sizes and temperatures to fully cook two identical large pots of spaghetti. Note that all the pots are as large as the largest hot plate. The shading of each hot plate is used to illustrate its temperature. The darker the shade of gray the cooler the temperature of the hot plate.

1) For each pair of hot plates shown below, circle the one that will cook the large pot of spaghetti more quickly. If there is no way to tell for sure, state that explicitly.

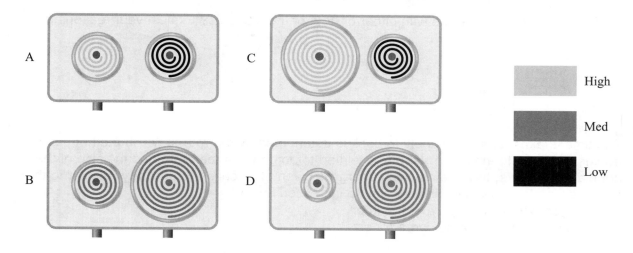

2) If you use two hot plates of the same size, can you assume that the hot plate that can cook a large pot of spaghetti first is at the higher temperature? Which lettered example above supports your answer?

3) If you use two hot plates at the same temperature, can you assume that the hot plate that can cook a large pot of spaghetti first is larger? Which lettered example above supports your answer?

4) If you use two hot plates of different sizes, can you assume that the hot plate that can cook a large pot of spaghetti first is at a higher temperature? Which lettered example above supports your answer?

5) Two students are discussing their answers to Question 4:

Student 1:   *In 1D, the hot plate on the left cooks the spaghetti quicker than the one on the right even though it is smaller. The hot plate's higher temperature is what makes it cook the spaghetti more quickly.*

Student 2:   *But the size of the hot plate also plays a part in making it cook fast. If the hot plate on the left were the size of a penny, the spaghetti would take a really long time to cook. I bet that if the size difference were great enough, the one at the lower temperature could cook the spaghetti first.*

Do you agree or disagree with either or both of the students? Explain your reasoning.

The time it takes for the spaghetti to cook is determined by the rate at which the hot plate transfers energy to the pot. This rate is related to both the temperature and the size of the hot plate. For stars, the rate at which energy is given off is called **luminosity.** Similar to the above example, a star's luminosity can be increased by
*   increasing its temperature; and/or
*   increasing its surface area (or size).

This relationship between luminosity, temperature, and size allows us to make comparisons between stars.

6) If two hot plates have the same temperature and one cooks the pot of spaghetti more quickly, what can you conclude about the sizes of the hot plates?

7) Likewise, if two stars have the same surface temperature and one is more luminous, what can you conclude about the sizes of the stars?

8) If two stars have the same surface temperature and are the same size, which star, if either, is more luminous? Explain your reasoning.

9) If two stars are the same size, but one has a higher surface temperature, which star, if either, is more luminous? Explain your reasoning.

Part II: Application to the H–R Diagram

The graph below plots the luminosity of a star on the vertical axis against the star's surface temperature on the horizontal axis. This type of graph is called an H–R diagram. Use the H–R diagram below and the relationship between a star's luminosity, temperature, and size (as described on the previous page) to answer the following questions concerning the stars labeled U–Y.

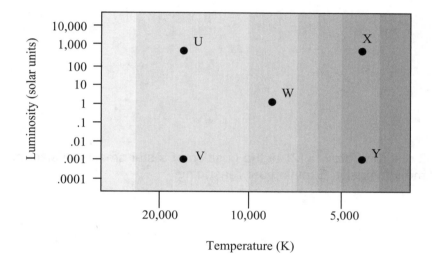

10) Stars U and V have the same surface temperature. Given that Star U is actually much more luminous than Star V, what can you conclude about the size of Star U compared to Star V? Explain your reasoning.

11) Star U has a greater surface temperature than Star X. Given that Star X is actually just as luminous as Star U, what can you conclude about the size of Star X compared to Star U? Explain your reasoning.

12) Based on the information presented in the H–R diagram, which star is larger, X or Y? Explain your reasoning.

13) Based on the information presented in the H–R diagram, which star is larger, Y or V? Explain your reasoning.

14) On the H–R diagram, draw a "Z" at the position of a star smaller in size than Star W but with the same luminosity. Explain your reasoning.

15) It is very difficult to accurately predict how the size of Star U will compare to that of Star W (without performing some kind of calculation). Explain what makes a comparison of the size of these stars so difficult.

Part I: Spectral Curves

White light is made up of all colors of light. We can see the individual colors when white light is passed through a prism or when we look at a rainbow. Light can come in an array of types or forms, which we call a *spectrum*. A *spectral curve* (like the one shown below) is a graph that displays the amount of energy given off by an object each second versus the different wavelengths (or colors) of light. For a specific color of light on the horizontal axis, the height of the curve will indicate how much energy is being given off at that particular wavelength. Figure 1 shows the spectral curve for an object emitting more red and orange light than indigo and violet. Notice that the red end of the curve is higher than the violet end, so the object will appear slightly reddish in color.

1) Which color of light has the greatest energy output in Figure 1?

2) Imagine that the blue light and orange light from the source were blocked. What color(s) would now be present in the spectrum of light observed?

3) Which of the following is the most accurate spectral curve for the spectrum described in Question 2?

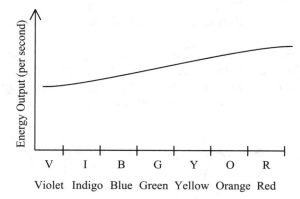

Figure 1

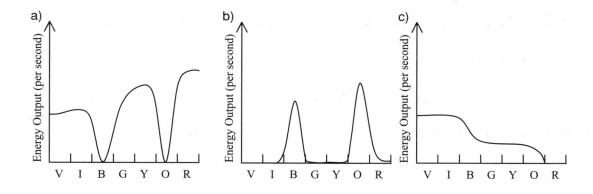

4) What colors of light are present in 3b above?

5) What colors are present in 3c above? Would this object appear reddish or bluish?

Part II: Blackbody Curves

Different colors of light are manifestations of the same phenomenon but have different wavelengths. For example, red light has a wavelength between 650 nm and 750 nm, whereas violet light has a shorter wavelength between 350 nm and 450 nm. Stars also give off light at wavelengths outside the visible part of the spectrum, as seen in Figures 2a, 2b, and 2c.

The two most important features of a star's spectral curve (also known as a blackbody curve) are:
*   its maximum height or peak—where the energy output is greatest; and
*   the corresponding wavelength at which this peak occurs—which indicates the star's temperature. If the peak occurs at a long wavelength, the star is cooler than a star that gives off most of its light (peaks) at a short wavelength.

For example, if Star E and Star F are the same size and temperature, they will have identical blackbody curves. However, if Star F is the same size as Star E, but is cooler, then its energy output is <u>less at all wavelengths</u> and the peak occurs <u>at a longer wavelength</u> (toward the red end of the spectrum).

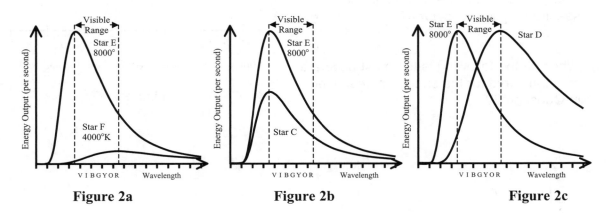

Figure 2a   **Figure 2b**   **Figure 2c**

Use Figure 2a to answer Questions 6–9. Assume Stars E and F are the same size.

6) Which star gives off more red light? Explain your reasoning.

7) Which star gives off more blue light? Explain your reasoning.

8) Which star looks redder? Explain your reasoning.

9) Two students are discussing their answers to Question 8.

Student 1: *Star E looks redder because it is giving off more red light than Star F.*
Student 2: *I disagree, you're ignoring how much blue light Star E gives off. Star E gives off more blue light than red light, so it looks bluish. Star F gives off more red than blue, so it looks reddish. That's why Star F looks redder than Star E.*

Do you agree or disagree with either or both of the students? Explain your reasoning.

10) Using the blackbody curves shown in Figure 2b, for each characteristic listed in the table below circle the correct response in the column to the right.

Characteristic	Responses			
Peaks at a longer wavelength	Star E	Star C	They peak at the same wavelength	
Has a lower surface temperature	Star E	Star C	They have the same surface temperature	
Looks red	Star E	Star C	They both look red	Neither looks red
Looks blue	Star E	Star C	They both look blue	Neither looks blue
Has a greater energy output	Star E	Star C	The have the same energy output	

11) How must Star C be different from Star E to account for their difference in energy output? Explain your reasoning.

12) Two students are discussing their answers to Question 11.

Student 1: *The peaks are at the same place so they must be at the same temperature. If Star C were as big as Star E, it would have the same output. Since the output is lower, Star C must be smaller.*
Student 2: *No. If its output is lower, it must be cooler. Since the temperatures of the two stars are the same, they must be the same size.*

Do you agree or disagree with either or both of the students? Explain your reasoning.

Consider the blackbody curves for Stars E and D shown in Figure 2c when answering Questions 13–15.

13) For each star, describe its color as either reddish or bluish.

Star E: Star D:

14) Which star has the greater surface temperature? Explain your reasoning.

15) Which star is larger? Explain your reasoning. (Hint: Consider how the energy output and temperatures for the two stars compare.)

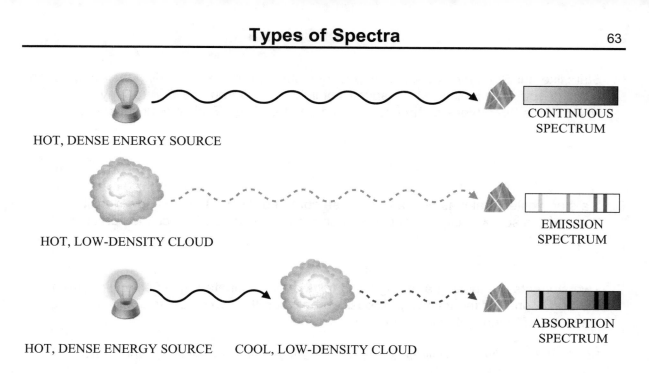

HOT, DENSE ENERGY SOURCE — CONTINUOUS SPECTRUM

HOT, LOW-DENSITY CLOUD — EMISSION SPECTRUM

HOT, DENSE ENERGY SOURCE   COOL, LOW-DENSITY CLOUD — ABSORPTION SPECTRUM

1) What type of spectrum is produced when the light emitted directly from a hot, dense object passes through a prism?

2) What type of spectrum is produced when the light emitted directly from a hot, low-density cloud of gas passes through a prism?

3) Describe in detail the source of light and the path the light must take to produce an absorption spectrum.

4) There are dark lines in the absorption spectrum that represent missing light. What happened to this light that is missing in the absorption line spectrum?

5) Stars like our Sun have low-density, gaseous atmospheres surrounding their hot, dense cores. If you were looking at the spectra of light coming from the Sun (or any star), which of the three types of spectrum would be observed? Explain your reasoning.

6) If a star existed that was only a hot, dense core and did **NOT** have a low-density atmosphere surrounding it, what type of spectrum would you expect this particular star to give off?

7) Two students are looking at a brightly lit full Moon, illuminated by reflected light from the Sun. Consider the following discussion between the two students about what the spectrum of moonlight would look like.

    Student 1:   *I think moonlight is just reflected sunlight, so we will see the Sun's absorption line spectrum.*

    Student 2:   *I disagree. An absorption spectrum has to come from a hot, dense object. Since the Moon is not a hot, dense object, it can't give off an absorption line spectrum.*

Do you agree or disagree with either or both of the students? Explain your reasoning.

8) Imagine that you are looking at two different spectra of the Sun. Spectrum #1 is obtained using a telescope that is in a high orbit far above Earth's atmosphere. Spectrum #2 is obtained using a telescope located on the surface of Earth. Label each spectrum below as either Spectrum #1 or Spectrum #2.

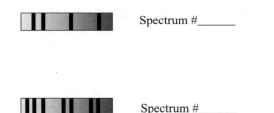

          Spectrum #_____

          Spectrum #_____

Explain the reasoning behind your choices.

In this activity, we will use a representation of the atom in which a central nucleus containing the protons and neutrons is surrounded by circles that represent the energy levels electrons can occupy.

1) Draw an atom including a nucleus and five energy levels that electrons could occupy. Use a dot to represent an electron at the lowest energy level.

One way an atom emits light (photons) occurs when an electron drops down from a high energy level (also referred to as an excited state) to a lower energy level (the lowest energy level is referred to as the ground state.)

2) Will an atom emit light if all of the atom's electrons are in the ground state? Explain your reasoning.

3) In which case does an atom emit more energy (*circle one*)?

Case A: *An electron drops down from the first excited state to the ground state.*
Case B: *An electron drops down from the third excited state to the ground state.*

Explain your reasoning.

4) Two students are talking about how light is emitted from atoms. Consider the following discussion between the two students and the sketches each student drew to illustrate their thinking.

Student 1:   *I drew my atom like this because my professor said that the gap between the energy levels gets bigger and bigger as you go up in energy from the ground state.*

Student 2:   *I think you've got it backward. The gap between energy levels will get smaller as you go up in energy levels, like I've drawn.*

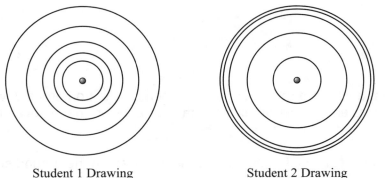

Student 1 Drawing   Student 2 Drawing

Do you agree or disagree with either or both of the students? Explain your reasoning.

5) A solid, glowing-hot object will emit light over the full range of wavelengths resulting in a continuous spectrum. If a diffuse and relatively cool cloud of gas is located between the glowing, hot object and an observer, what type of spectrum will the observer detect coming out of the cloud (*circle one*)?

continuous spectrum   absorption spectrum   emission spectrum

Explain the reasoning behind your choice.

6) At the right is a sketch showing one of the atoms in the diffuse, cool cloud of gas described in the previous question. Note that the atom has several energy levels that an electron could exist in. Using a dot to represent an electron, a straight arrow to represent the motion of the electron, and a squiggly arrow to represent the photon, sketch what you think would happen within this atom to cause the type of spectrum described in the previous question. Explain the reasoning behind why you drew the electron and arrows the way you did.

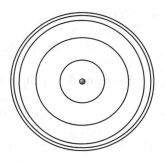

7) Imagine that you are looking at a neon sign in a store window that says "OPEN." This sign can be thought of as a tube filled with a gas of neon atoms that have electrons changing from one energy state to a different energy state and in the process are giving off mostly red light. Which type of spectrum would you observe coming from the "OPEN" sign (*circle one*)?

continuous spectrum   absorption spectrum   emission spectrum

Explain the reasoning behind your choice.

8) At the right is a sketch showing one of the atoms in the neon sign described in the previous question. Note that the atom has several energy levels that an electron could exist in. Using a dot to represent an electron, a straight arrow to represent the motion of the electron, and a squiggly arrow to represent the photon, sketch what you think would happen within this atom to cause the type of spectrum described in the previous question. Explain the reasoning behind why you drew the electron and arrows the way you did.

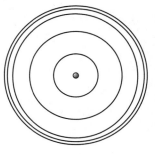

LECTURE-TUTORIALS FOR INTRODUCTORY ASTRONOMY
THIRD EDITION

9) Consider the following discussion between the two students about the atoms and light coming from the red "OPEN" sign from the previous question and the light coming from a yellow "OPEN" sign they see across the street.

Student 1:   *I think that all you need to do to get signs to give off light of different colors is to use tinted glass of different colors. If you electrify a gas, white light is emitted so the color of glass gives the sign its color.*

Student 2:   *That can't be it because the electrons will always move between the same energy levels for atoms in a neon sign and so they will mostly give off red light. I think the yellow sign is filled with a different type of atoms with energy levels that are farther apart, so when the electrons drop down from a higher energy level to a lower energy level, the atoms will give off yellow light instead of red light.*

Do you agree or disagree with either or both of the students? Explain your reasoning.

10) Redraw the initial drawing you made in Question 1. Describe what additions or changes you made on this new drawing so that it better conveys what you understand about the relationship between light and atoms.

11) Use the hypothetical atom drawings (A–F) below to answer the next five questions. Note there is only one correct choice for each question and each choice is used only once.

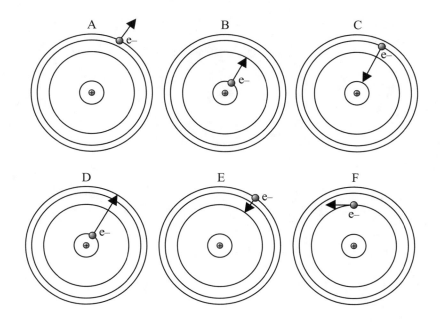

a) Which shows the absorption of violet light? Explain your reasoning.

b) Which shows the emission of blue light? Explain your reasoning.

c) Which shows the absorption of green light? Explain your reasoning.

d) Which shows the emission of orange light? Explain your reasoning.

e) Which shows an electron being ejected from the atom?

The absorption line spectra for six hypothetical stars, each with different temperatures, are shown below. For each absorption line spectrum, the short wavelengths of light (or blue end) of the electromagnetic spectrum are shown on the left side, and the long wavelengths of light (or red end) of the spectrum are shown on the right side.

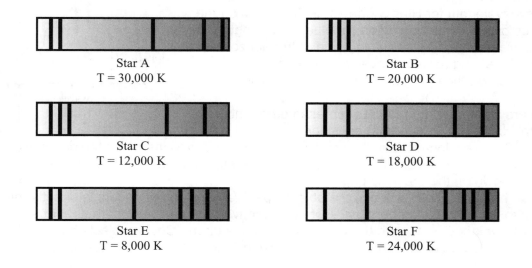

Star A
T = 30,000 K

Star B
T = 20,000 K

Star C
T = 12,000 K

Star D
T = 18,000 K

Star E
T = 8,000 K

Star F
T = 24,000 K

1) Do cold stars always appear to have a different (greater or fewer) number of lines in their absorption spectra than hot stars? Cite evidence from the above spectra to support your answer.

2) Do cold stars always appear to have more lines at either the blue or red ends of their absorption spectra than hot stars? Cite evidence from the above spectra to support your answer.

3) Consider the absorption line spectrum given below for Star G. Can you determine the approximate temperature for Star G by comparing its absorption line spectrum to the absorption line spectra and temperatures of Stars A–F given above? If so, write in your estimate in the space below; if not, explain why not.

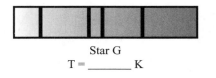

Star G
T = _____ K

While it is difficult to directly estimate the temperature of a star from the lines in its absorption spectrum, we can always use the wavelength of the peak for the light emitted by a star to estimate temperature. The spectral curve on the graph at right illustrates the energy output versus wavelength for Star G (the same Star G as on the previous page). Again, the short (or blue) wavelengths of light are represented on the left side of the horizontal axis, and the long (or red) wavelengths of light are represented on the right end of the horizontal axis.

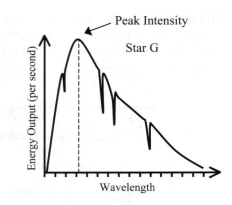

There are two important features represented on this spectral curve that you need to consider.

- The exact locations of the small dips, or absorption features, on the curve occur at the same wavelength as the dark lines that appear in the absorption line spectrum shown on the previous page.
- The wavelength at which the object's peak energy output occurs is directly related to the object's temperature. Hot objects have their peak energy output at short wavelengths—toward the blue end of the spectrum. Cooler objects have their peak energy output at long wavelengths—toward the red end of the spectrum.

Although the total energy output of a star is affected by its temperature (and, therefore, so is the height of the spectral curve), for this activity, we will assume that the height (but not location) of the peak energy output, and the general shape of the spectral curves for Stars A–F, can be drawn nearly the same for each star. Only the location of the peak intensity and the location of the small dips, or absorption features, will be different for each star.

4) Examine the spectral curve shown at right for Star A (the same Star A as on the previous page). Note that Star G has a temperature of approximately 25,000 K, whereas Star A has a temperature of 30,000 K. With this information, would you say the location for the peak of Star A is drawn at approximately the correct wavelength as compared to the wavelength where the peak of Star G is drawn? Explain your reasoning.

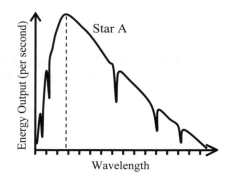

5) Are the absorption features (dips) in the spectral curve for Star A drawn at approximately the correct wavelengths? Explain how you can tell.

6) Sketch spectral curves for Stars B–F on the corresponding graphs provided below. Do not worry about whether the heights of the spectral curves you draw are accurate. But, make sure your spectral curves include the absorption features and the peak intensities drawn at *approximately* the correct wavelengths.

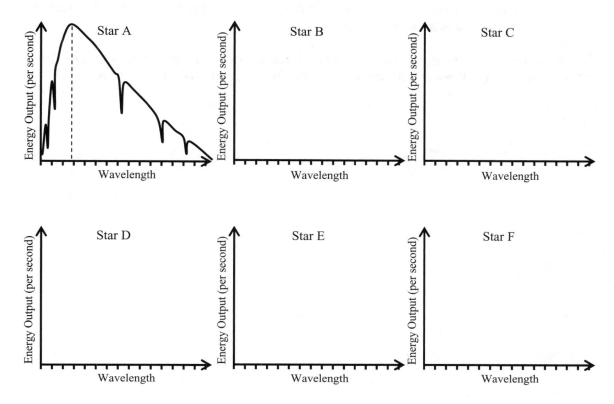

7) Did you draw the peak intensity of each spectral curve at the same wavelength as the spectral curve for Star A? Why or why not?

8) If you were given a star's absorption line spectrum and its corresponding spectral curve shown on an energy output per second versus wavelength graph, how could you approximate the temperature of the star?

9) Consider the following statement made by a student regarding a star's temperature and its corresponding absorption line spectrum.

Student: *If I am looking at a star's absorption line spectrum and see that it has a lot of lines at the blue end of the spectrum, then the star must be hot because the blue lines are higher energy lines.*

Do you agree or disagree with this student? Explain your reasoning and support your answer by citing evidence from the absorption line spectra given for Stars A–G.

Because of the Doppler effect, light emitted by an object can appear to change wavelength due to its motion toward or away from an observer. When the observer and the source of light are moving toward each other, the light is shifted to shorter wavelengths (blueshifted). When the observer and the source of light are moving away from each other, the light is shifted to longer wavelengths (redshifted).

Part I: Motion of Source

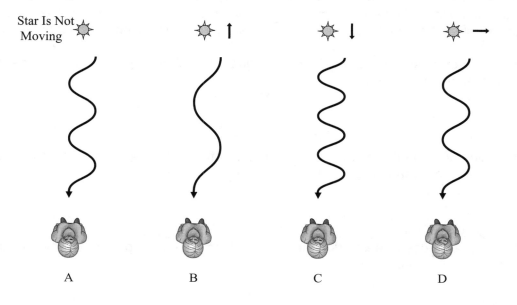

1) Consider the situations shown (A–D).

   a) In which situation will the observer receive light that is shifted to shorter wavelengths?

   b) Will this light be blueshifted or redshifted for this case?

   c) What direction is the star moving relative to the observer for this case?

2) Consider the situations shown (A–D).

   a) In which situation will the observer receive light that is shifted to longer wavelengths?

   b) Will this light be blueshifted or redshifted for this case?

   c) What direction is the star moving relative to the observer for this case?

3) In which of the situations shown (A–D) will the observer receive light that is not Doppler shifted at all? Explain your reasoning.

4) Imagine our solar system is moving in the Milky Way toward a group of three stars. Star A is a blue star that is slightly closer to us than the other two. Star B is a red star that is farthest away from us. Star C is a yellow star that is halfway between Stars A and B.

   a) Which of these three stars, if any, will give off light that appears to be blueshifted? Explain your reasoning.

   b) Which of these three stars, if any, will give off light that appears to be redshifted? Explain your reasoning.

   c) Which of these three stars, if any, will give off light that appears to have no shift? Explain your reasoning.

5) You overhear two students discussing the topic of Doppler shift.

   Student 1:   *Since Betelgeuse is a red star, it must be going away from us, and since Rigel is a blue star it must be coming toward us.*
   Student 2:   *I disagree, the color of the star does not tell you if it is moving. You have to look at the shift in wavelength of the lines in the star's absorption spectrum to determine whether it's moving toward or away from you.*

   Do you agree or disagree with either or both of the students? Explain your reasoning.

Part II: Shift in Absorption Spectra

When we study an astronomical object like a star or galaxy, we examine the spectrum of light it gives off. Since the lines of a spectrum occur at specific wavelengths, we can determine that an object is moving when we see that the lines have been shifted to either longer or shorter wavelengths. For the absorption line spectra shown on the next page, short-wavelength light (the blue end of the spectrum) is shown on the left-hand side, and long-wavelength light (the red end of the spectrum) is shown on the right-hand side.

For the three absorption line spectra shown below (A, B, and C), one of the spectra corresponds to a star that is not moving relative to you, one of the spectra is from a star that is moving toward you, and one of the spectra is from a star that is moving away from you.

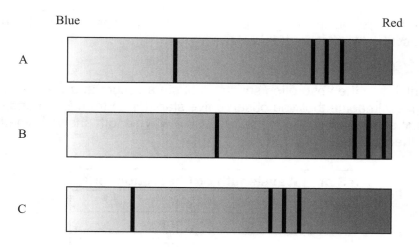

6) Which of the three spectra above corresponds with the star moving toward you? Explain your reasoning.

7) Which of the three spectra corresponds with the star moving away from you? Explain your reasoning.

Part III: Size of Shift and Speed

If two sources of light are moving relative to an observer, the light from the star that is moving faster will appear to undergo a greater Doppler shift.

Consider the four spectra at the right. The spectrum labeled F is an absorption line spectrum from a star that is at rest. Again, note that short-wavelength (blue) light is shown on the left-hand side of each spectrum, and long-wavelength (red) light is shown on the right-hand side of each spectrum.

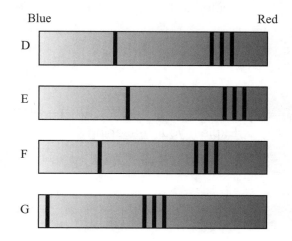

8) Which of the four spectra would be from the star that is moving the fastest? Would this star be moving toward or away from the observer?

9) Of the stars that are moving, which spectra would be from the star that is moving the slowest? Describe the motion of this star.

10) An important line in the absorption spectrum of stars occurs at a wavelength of 656 nm for stars at rest. Imagine that you observe five stars (H–L) from Earth and discover that this important absorption line is measured at the wavelength shown in the table below for each of the five stars.

Star	Wavelength of Absorption Line
H	649 nm
I	660 nm
J	656 nm
K	658 nm
L	647 nm

a) Which of the stars are giving off light that appears blueshifted? Explain your reasoning.

b) Which of the stars are giving off light that appears redshifted? Explain your reasoning.

c) Which star is giving off light that appears shifted by the greatest amount? Is this light shifted to longer or shorter wavelengths? Explain your reasoning.

d) Which star is moving the fastest? Is it moving toward or away from the observer? Explain your reasoning.

e) Which of the stars (H–L) would appear blue? Which of the stars (H–L) would appear red? Explain your reasoning.

f) Which of the stars (H–L) is closest to Earth? Which of the stars (H–L) is furthest from Earth? Explain your reasoning. If you cannot determine which star (H–L) is closest or furthest from Earth, explain why not.

11) The figure at right shows a spaceprobe and five planets (A-E). The motion of the spaceprobe is indicated by the arrow. The spaceprobe is continuously broadcasting a radio signal in all directions.

a) Which planets will receive a radio signal that is redshifted? Explain your reasoning.

A

C

D

b) Which planets will receive a radio signal that is shifted to shorter wavelengths? Explain your reasoning.

B

E

c) Will all the planets receive radio signals from the spaceprobe that are Doppler shifted? Explain your reasoning.

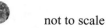
not to scale

d) How will the size of the Doppler shift in the radio signals detected at Planets A and B compare? Explain your reasoning.

e) How will the size of the Doppler shift in the radio signals detected at Planets E and B compare? Explain your reasoning.

Figure 1 shows Earth, the Sun, and five different possible positions for the Moon during one full orbit (dotted line). It is important to recall that one-half of the Moon's surface is illuminated by sunlight at all times. For each of the five positions of the Moon shown below, the Moon has been shaded on one side to indicate the half of the Moon's surface that is **not** being illuminated by sunlight. Note that this drawing is not to scale.

1) Which Moon position (A–E) best corresponds with the Moon phase shown in the upper-right corner of Figure 1? Make sure that the Moon position you choose correctly predicts a Moon phase in which only a small crescent of light on the left-hand side of the Moon is visible from Earth.

   Enter the letter of your choice: _____

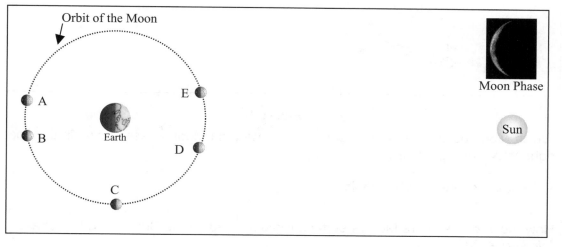

Figure 1

2) In the blank boxes below, sketch how the Moon would appear from Earth for the four Moon positions that you did **not** choose in Question 1. Be sure to label each sketch with the corresponding letter indicating the Moon's position from Figure 1.

3) Shade in each of the four Moons shown in Figure 2 to indicate which portion of the Moon's surface will **not** be illuminated by sunlight.

Use Figure 2 to answer Questions 4–7.

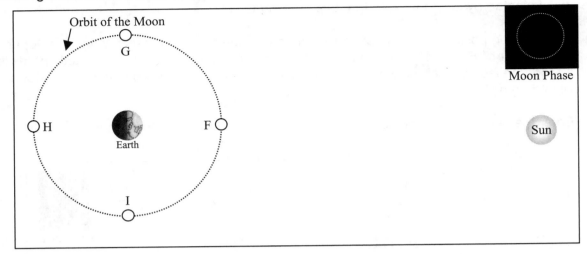

Figure 2

4) Which Moon position (F–I) best corresponds with the Moon phase shown in the upper-right corner of Figure 2?

   Enter the letter of your choice: _____

5) How much of the entire Moon's surface is illuminated by the Sun during this phase (*circle one*)?

   a) None of the surface is illuminated.

   b) Less than half of the surface is illuminated.

   c) Half of the surface is illuminated.

   d) More than half of the surface is illuminated.

   e) All of the surface is illuminated.

6) How much of the Moon's illuminated surface is visible from Earth for this phase of the Moon (*circle one*)?

   a) None of the surface (visible from Earth) is illuminated.

   b) Less than half of the surface (visible from Earth) is illuminated.

   c) Half of the surface (visible from Earth) is illuminated.

   d) More than half of the surface (visible from Earth) is illuminated.

   e) All of the surface (visible from Earth) is illuminated.

7) Would your answers to Questions 5 or 6 change if the Moon were in the third-quarter phase rather than the phase shown in Figure 2? Explain your reasoning.

8) Consider the following discussion between two students about the cause of the phases of the Moon.

Student 1: *The phase of the Moon depends on how the Moon, Sun, and Earth are aligned with one another. During some alignments only a small portion of the Moon's surface will receive light from the Sun, in which case we would see a crescent Moon.*

Student 2: *I disagree. The Moon would always get the same amount of sunlight; it's just that in some alignments Earth casts a larger shadow on the Moon. That's why the Moon isn't always a full Moon.*

Do you agree or disagree with either or both of the students? Explain your reasoning.

1) If the Moon is a full Moon tonight, will the Moon be waxing or waning one week later? Which side of the Moon (right or left) will appear illuminated at this time?

   Circle one: Waxing or Waning

   Circle one: Right or Left

2) Where (in the southern sky, on the eastern horizon, on the western horizon, high in the sky, etc.) would you look to see the full Moon when it starts to rise? What time would this happen?

3) Where (in the southern sky, on the eastern horizon, on the western horizon, high in the sky, etc.) would you look to see the Sun when the full Moon starts to rise?

4) Where (in the southern sky, on the eastern horizon, on the western horizon, high in the sky, etc.) would you look to see the new Moon, if it were visible, when it starts to rise? What time would this happen?

5) If the Moon is a new Moon when it rises, which of the phases shown below (A–H) will it be in when it sets?

   Letter of Moon phase: _____

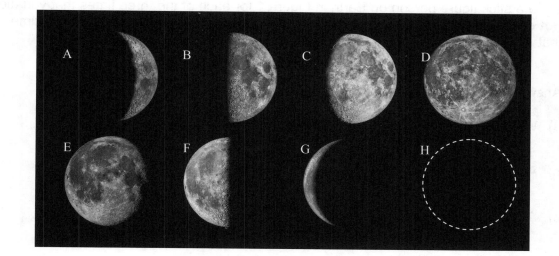

LECTURE-TUTORIALS FOR INTRODUCTORY ASTRONOMY
                                                                        THIRD EDITION

Figure 1 shows the position of the Sun, Earth, and Moon for a particular phase of the Moon. The Moon has been shaded on one side to indicate the portion of the Moon that is **not** being illuminated by sunlight. A person has been placed on Earth to indicate an observer's position at noon. Recall that with this representation Earth will complete one counterclockwise rotation in each day. Note that this drawing is not to scale.

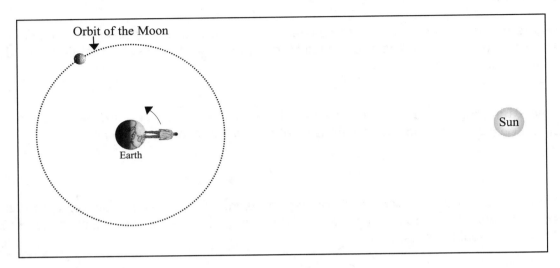

Figure 1

6) What time is it for the person shown in Figure 1?

   Circle one: 6 A.M (sunrise)   12 P.M. (noon)   6 P.M. (sunset)   12 A.M (midnight)

7) Draw a stick figure person on Earth in Figure 1 for each of the three times that you did **not** choose in Question 6. Label each of the stick figures that you drew with the time that the person would be located there.

8) Answer the following questions for the position of the Moon shown in Figure 1.

   a) Which Moon phase would an Earth observer see?

   b) At what time will the Moon shown appear highest in the sky?

   c) At what time will the Moon shown appear to rise?

   d) At what time will the Moon shown appear to set?

9) At what time would you look to see a first-quarter Moon at its highest position in the sky?

10) If the Sun set below your western horizon about 2 hours ago, and the Moon is barely visible on the eastern horizon, what phase would the Moon be in at this time and location?

11) A friend comments to you that there was a beautiful, thin sliver of Moon visible in the early morning just before sunrise. Which phase of the Moon would this be, and in what direction would you look to see the Moon (in the southern sky, on the eastern horizon, on the western horizon, high in the sky, etc.)? Draw and label a picture that explains your reasoning.

Figure1 illustrates the sky as seen from the continental United States. It shows that the Sun's daily path across the sky (dashed/solid line) is longest on June 21 and shortest on December 21. In addition, on June 21, which is called the summer solstice, the Sun reaches its maximum height in the southern sky above the horizon at about noon. The figure shows that the Sun never actually reaches the zenith for any observer in the continental United States. In other words, the Sun is never directly overhead. Over the six months following the summer solstice, the height of the Sun at noon moves progressively lower and lower until December 21, the winter solstice. Thus, we see that the path of the Sun through the southern sky changes considerably over the course of a year.

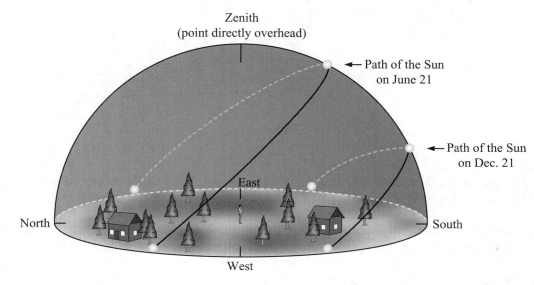

Figure 1

1) According to Figure 1, in which direction would you look to see the Sun when it reaches its highest position in the sky today?

   Circle one: east   southeast   south   southwest   west

2) If it is wintertime right now (just after the winter solstice), how does the height of the Sun at noon change over the next several months?

   Circle one: increases   stays the same   decreases

3) Since Figure 1 is a reasonable representation for observers in the continental United States, is there ever a time of year when the Sun is directly overhead at the zenith (looking straight up) at noon? If so, on what date does this occur?

4) During what time(s) of year would the Sun rise:

   a) north of east?

   b) south of east?

   c) directly in the east?

LECTURE-TUTORIALS FOR INTRODUCTORY ASTRONOMY
                                                              THIRD EDITION

5) Does the Sun always set in precisely the same location throughout the year? If not, describe in what way the direction of where the Sun sets changes throughout the year.

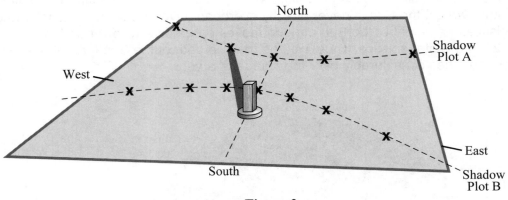

Figure 2

Figure 2 shows a small, vertical stick, which casts a shadow while it rests on a large piece of paper or poster board. You can think of this to be somewhat like a sundial.

For two different days of the year, the end of the shadow has been marked with an **x** every couple of hours throughout the day. Although this sketch is somewhat exaggerated, these *shadow plots* indicate how the position of the Sun changes in the sky through the course of these two days. The following questions are designed to show the relationship between Figure 1 on the previous page and Figure 2 above.

6) What do the **x**'s in the shadow plots represent?

7) Approximately how much time went by from the time one of the **x**'s was drawn until the next **x** was drawn for each shadow plot?

8) Approximately how long did it take to create each of the shadow plots?

9) How does the direction of the stick's shadow compare to the location of the Sun at the time each **x** was drawn?

10) Using Figures 1 and 2, in what direction would the shadow of the stick be cast on the poster board if the Sun rises in the southeast?

   Circle one: west   northwest   north   northeast   east   southeast

11) Clearly circle the **x** for the shadow that corresponds to noon for Shadow Plot A and for Shadow Plot B.

12) Compare the position of the **x** that corresponds to noon for Shadow Plots A and B. Which Shadow Plot (A or B) corresponds to a path of the Sun in which the Sun is highest in the sky at noon? Explain your reasoning.

13) Which Shadow Plot (A or B) most closely corresponds to the Sun's path through the sky during the summer, and which corresponds with the winter? Label these paths on Figure 2. Explain your reasoning.

14) On Figure 2, sketch the Sun's position shortly after sunrise in the summer and label the **x** that indicates the position of the end of the stick's shadow at this time. Explain your reasoning for why you sketched the Sun where you did and labeled the **x** that you did.

15) Based on the shadow plots in Figure 2, during which time of the year (summer or winter) does the Sun rise to the south of east? Explain your reasoning.

16) If Shadow Plot A corresponds to the path of the Sun on the day of the winter solstice, is it possible that there would ever be a time when the stick would cast a shadow longer than the one shown along the north-to-south line that indicates the Sun's position at noon? Explain your reasoning.

17) If Shadow Plot B corresponds to the path of the Sun on the day of the summer solstice, is it possible that there would ever be a time when the stick would cast a shadow shorter than the one shown along the north-to-south line that indicates the Sun's position at noon? Explain your reasoning.

18) If you were to mark the end of the stick's shadow with an **x**, where would the **x** be placed along the north-to-south line to indicate the Sun's position at noon *today*? Clearly explain why you placed the **x** where you did.

19) Will the stick ever cast a shadow along the north-to-south line that extends to the south of the stick at noon? Explain your reasoning.

20) Is there ever a clear (no clouds) day of the year in the continental United States when the stick casts no shadow? If so, when does this occur, and where exactly in the sky does the Sun have to be?

Part I: Earth–Sun Distance

Listed below are the distances, in kilometers (km), between the Sun and Earth for four months of the year. The drawing at the right shows four different locations of Earth during its orbit around the Sun. Note that for each location drawn, Earth is correctly shown with its rotational axis tilted at an angle of 23.5°.

Drawing not to scale

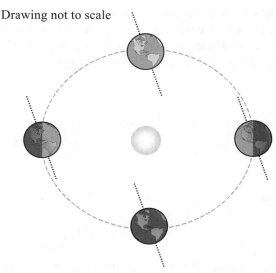

Month	Earth–Sun Distance
December	147.2 million km
June	152.0 million km
September	150.2 million km
March	149.0 million km

1) Is the direction that Earth's axis is tilted changing as Earth orbits the Sun?

2) Using the information listed above, does Earth stay the same distance from the Sun throughout the year? If not, what month(s) and during which season (for the Northern Hemisphere) is Earth closest to the Sun? Farthest from the Sun?

3) Would you say the temperature stays approximately the same every month of the year at your location?

4) Are the seasons (summer or winter) the same in the Northern and Southern Hemispheres at the same time? When it is summer in the Northern Hemisphere, what season is it in the Southern Hemisphere?

5) Consider the following discussion between two students about the cause of the seasons.

Student 1: *I know that it's hotter in the summer and colder in the winter, so we must be closer to the Sun in the summer than in the winter.*

Student 2: *I disagree. Although the distance between Earth and the Sun does change throughout the year, I don't believe that the seasons and changes in Earth's surface temperature are caused by the distance between the Sun and Earth. If the seasons were due to the Sun–Earth distance, then both hemispheres of Earth would have the same seasons at the same time.*

Do you agree or disagree with either or both of the students? Explain your reasoning.

At different times of the year, locations in the Northern Hemisphere can be a few thousand kilometers closer to (or farther from) the Sun than locations that are at the same latitude in the Southern Hemisphere (as shown in the drawing below). However, the distance between Earth and the Sun is, on average, about 150 million kilometers.

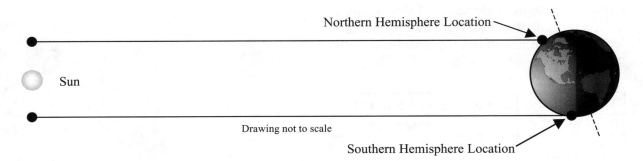

6) Do you think these differences in distance between locations at the same latitude in the Northern and Southern Hemispheres are the cause of the seasons? Explain your reasoning.

7) Consider the following discussion between two students about the cause of the seasons.

Student 1:   *I get it. So since Earth is tilted, there are times when the northern part of Earth is closer to the Sun than the southern part. So the north has summer and the south has winter. And then later, the south is tilted toward the Sun and gets closer and has summer.*

Student 2:   *I disagree. Although the tilt does bring one hemisphere closer to the Sun, the difference in distance between the northern half and southern half of Earth is really small compared to how far away Earth is from the Sun.*

Do you agree or disagree with either or both of the students? Explain your reasoning.

Part II: Direct Light and Tilt

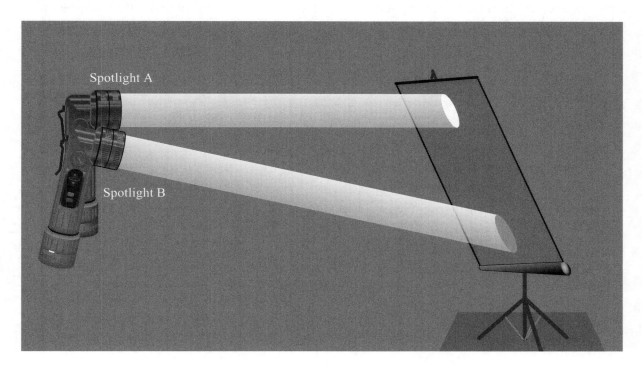

Consider the picture above in which two spotlights (A and B) are shown casting light onto a screen. Note: Each spotlight gives off the same total amount of light.

8) Which of the two lighted areas (the one created by Spotlight A or B) would appear brighter?

9) Which of the two lighted areas is smaller?

10) Which of the two lighted areas receives more direct light (amount of energy on each unit of area) from the spotlight?

11) If a thermometer were placed in each of the lighted areas, which one would read the higher temperature?

12) Which of the two positions would be similar to the way the sunlight would shine on the Southern Hemisphere of Earth during winter in the Southern Hemisphere? Explain your reasoning.

Consider the picture below illustrating three different regions of Earth (the Northern Hemisphere, the Southern Hemisphere, and the equatorial region) at two different times of the year, six months apart.

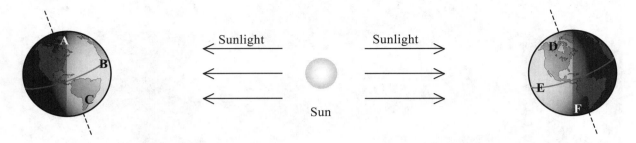

Note: this drawing is not to scale. In fact you could fit more than 100 Suns between the Sun and Earth.

13) Which location(s) (A–F) correspond(s) with summer in the Northern Hemisphere? Explain your reasoning.

14) Which location(s) (A–F) correspond(s) with winter in the Southern Hemisphere? Explain your reasoning.

Part III: Amount of Daylight

15) During which season (summer or winter) is the number of daylight hours the greatest? How many hours?

16) During which season (summer or winter) is the Sun highest in the sky at noon? Hint: Consider the drawing showing the lighted areas and the spotlights for Questions 8–12.

17) How are your answers to the previous two questions related to the time of year that your location experiences the highest average temperature? Explain your reasoning.

18) How would the number of hours of sunlight and the height of the Sun in the sky at noon change (if at all) over the course of the year for locations on the equator? Explain your reasoning.

IV: Applying the Model of Causes of Seasons

19) If, somehow, the number of daylight hours did not change throughout the year, but Earth was still tilted at 23.5°, would there still be seasons in the Northern and Southern Hemispheres of Earth? Would the temperature difference between the seasons still be as great? Explain your reasoning.

20) If the Northern Hemisphere were tilted 90° toward the Sun, which location would be warmer in summer: the Arctic Circle or Florida? Why?

21) Provide two pieces of evidence to support the fact that the varying distance between the Sun and Earth cannot account for the seasons.

22) Which two things are most directly responsible for the cause of the seasons on Earth? Explain your reasoning.

One of the most difficult parts of constructing an accurate model for planetary motions is that planets seem to wander among the stars. During their normal (or prograde) motion, planets appear to move from west to east over many consecutive nights as seen against the background stars. However, they occasionally (and predictably) appear to reverse direction and move east to west over consecutive nights as seen against the background stars. This backward motion is called retrograde motion.

1) Given the data in Table 1, plot the motion of the mystery planet on the graph provided in Figure 1 (record dates next to each position you plot). Then, draw a smooth line (or curve), using your data points, to illustrate the path of the planet through the sky.

Table 1 Mystery Planet Positions

Date of Observation	Azimuth (horizontal direction)	Altitude (vertical direction)
May 1	240	45
May 15	210	50
June 1	170	50
June 15	150	45
July 1	170	40
July 15	180	45
August 1	140	50
August 15	120	55

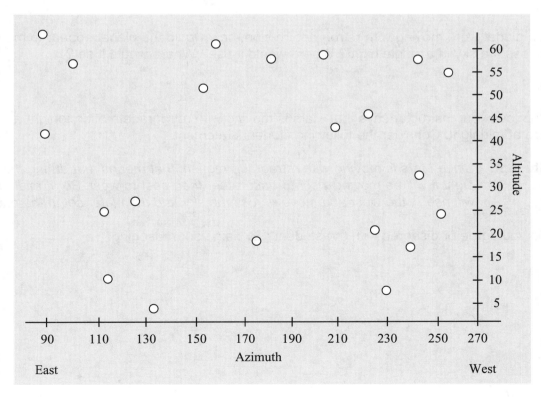

Figure 1

2) On what date was the mystery planet located farthest to the west? What was the azimuth value of the planet on this date?

3) On what date was the mystery planet located farthest to the east? What was the azimuth value of the planet on this date?

4) Describe how the mystery planet moved (east or west), as compared to the background stars, during the time between the dates identified in Questions 2 and 3.

5) During which dates does the mystery planet appear to move with normal, prograde, motion, as compared to the background stars? In what direction (east-to-west or west-to-east) does the planet appear to be moving relative to the background stars during this time?

6) During which dates does this mystery planet appear to move with backward, retrograde, motion, as compared to the background stars? In what direction (east-to-west or west-to-east) does the planet appear to be moving relative to the background stars during this time?

7) If a planet were moving with retrograde motion, how would the planet appear to move across the sky in a single night? Where would it rise? Where would it set?

8) Suppose your instructor says that Mars is moving with retrograde motion tonight and will rise at midnight. Consider the following student statement:

Student: *Since Mars is moving with retrograde motion, that means that during the night it will be moving west-to-east rather than east-to-west. So at midnight it will rise in the west and move across the sky and then later set in the east.*

Do you agree or disagree with the student? Explain your reasoning.

The extremely high temperature of Earth's core causes material in the surrounding mantle to become hot, expand, and rise toward the surface. The mantle material then cools and sinks, resulting in a circular motion of material moving beneath Earth's surface. This circulation of mantle material causes the continental and oceanic plates to move across Earth's surface. At various locations on Earth's surface, we are able to observe plates colliding, plates separating, and plates moving horizontally with respect to each other.

The drawing below shows a cross-section of Earth's surface and its underlying mantle. At this particular location of the surface, the dense oceanic plate is being forced beneath the less dense continental crust. The dense oceanic plate experiences higher temperatures (and pressures) as it is forced deeper into the mantle. This interaction between the oceanic plate and continental plate causes molten material to move upward through the continental plate until it breaks the surface in the form of volcanoes.

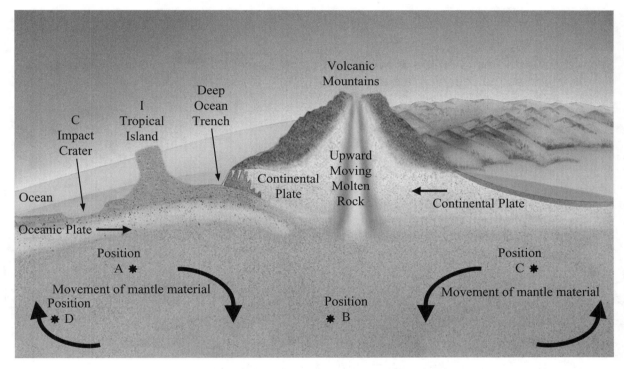

Oceanic to Continental Plate Convergence Zone

In the drawing above, Positions A–D show the position of four pieces of mantle material. Use this drawing to answer the next three questions.

1) Which direction (right or left) are the oceanic and continental plates moving?

2) Which is hotter, the piece of mantle material at Position A or the piece of mantle material at Position D? Explain your reasoning.

3) What direction are the pieces of mantle material moving (up, down, left, or right) at Positions A, B, C, and D?

4) Consider the following discussion between two students debating why the oceanic and continental plates move.

Student 1: *The plates are moving because the mantle material is constantly moving beneath Earth's plates, and this causes the plates to move.*

Student 2: *I disagree. The plates are just floating on the mantle material. The plates started moving a long time ago when Earth initially formed, and the plates' momentum keeps them moving toward each other.*

Do you agree or disagree with either or both of the students? Explain your reasoning.

5) Just beneath Point I on the drawing is a tropical island. What will eventually happen to the island as the oceanic plate moves? Why?

6) Just beneath Point C on the drawing is an ancient impact crater on the ocean floor where a giant comet collided with Earth. What will happen to the ancient impact crater as the oceanic plate moves? Why?

7) Imagine that an impact occurred on the continental plate millions and millions of years ago, leaving behind an impact crater near the right side of the base of the volcano. Why would there be little evidence of this impact crater found today?

8) Consider the image below of the rocky and crater-covered Moon. Its very old surface has remained virtually unchanged over the last few billion years. Do you think the Moon has an active, hot, and molten interior or an inactive, cold, and solid interior? Why?

9) If a new planet were discovered, what evidence would you look for to determine whether or not it has an active, hot, and molten interior? Why?

Objects give off different amounts of light depending upon their temperature. Figure 1, below, shows the energy output of our Sun along with the percent of energy given off by the Sun in the ultraviolet (UV), visible (VIS), and infrared (IR) portions of the electromagnetic spectrum.

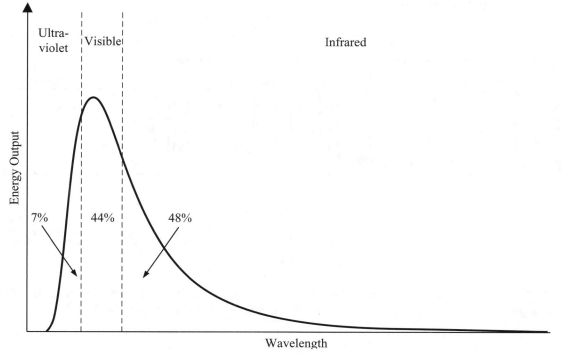

Figure 1

1) Which **TWO** forms of light account for the majority of energy coming from the Sun: ultraviolet, visible, or infrared? Which of the three accounts for the least energy? Provide numbers to support your answer.

2) Consider the following debate between two students regarding the energy given off by the Sun.

   Student 1: *I think that the Sun gives off most of its energy at ultraviolet wavelengths because ultraviolet light is more intense than visible light and you always hear about ultraviolet light causing sunburns.*

   Student 2: *Even though UV photons are more energetic than visible photons, the Sun simply gives off fewer ultraviolet photons and gives off way more visible and infrared photons. So I think that these longer wavelength photons account for most of the energy coming from the Sun.*

   Do you agree or disagree with either or both of these students? Explain your reasoning.

Earth's surface temperature is affected by light that is absorbed at the surface. However, a photon's ability to travel through our atmosphere depends upon its wavelength. Figure 2 below shows that some wavelengths of light are absorbed in our atmosphere more than others. The figure also lists the primary gas molecules responsible for absorbing the different wavelengths of light.

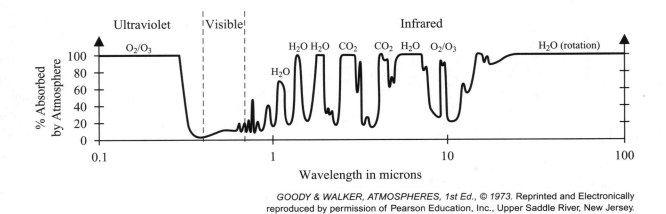

GOODY & WALKER, ATMOSPHERES, 1st Ed., © 1973. Reprinted and Electronically reproduced by permission of Pearson Education, Inc., Upper Saddle River, New Jersey.

Figure 2

3) Comparing the visible and the infrared types of light, which would you say has an easier time getting through our atmosphere? Which experiences more absorption?

4) Comparing the ultraviolet and the infrared types of light, which would you say has an easier time getting through our atmosphere? Which experiences more absorption?

5) Based upon Figures 1 and 2, why is ultraviolet light **NOT** an important energy source for heating the surface of Earth?

6) What gas molecules are primarily responsible for the absorption of each of the following types of light in our atmosphere?

Type of Light	Molecule(s) Responsible for Absorption
Ultraviolet	
Visible	
Infrared	

Molecules that are transparent to visible light but absorb and re-emit infrared light are known as "*greenhouse gases.*"

7) What are the two greenhouse gases most responsible for absorbing infrared light in Earth's atmosphere?

Once visible light from the Sun reaches the surface of Earth, some of the light is reflected back toward space as visible light, and the remaining light is absorbed by the ground. Reflected light does not change the temperature of the surface, whereas absorbed light causes the temperature of the surface to increase. Earth's heated surface then gives off infrared light to Earth's atmosphere. As an example, on a hot day, black asphalt absorbs more visible light and gives off more infrared light than does a white crosswalk.

8) The Sun is approximately 6000 K at the surface and has an energy distribution that peaks at visible wavelengths; Earth's surface is much cooler at about 288 K. What type of light do you think Earth's surface primarily gives off: ultraviolet, visible, or infrared light? Explain your reasoning.

9) Does Earth's surface give off light at night? If so, what type? If not, why not?

10) Consider the following debate between two students regarding the energy given off by Earth's surface.

Student #1: *The Sun mainly gives off visible light and so does Earth's surface because I can see it during the daytime.*

Student #2: *But that's just reflected sunlight. Earth's surface is much cooler than the Sun and mostly gives off energy closer to the kind that our bodies give off — infrared light. I'm not sure, but I think the surface probably radiates infrared light during both the daytime and the nighttime based upon its temperature.*

Do you agree or disagree with either or both of the students? Explain your reasoning.

11) Will the light given off by Earth's surface easily travel back through the atmosphere to space or will it be absorbed by molecules in the atmosphere? Explain your reasoning.

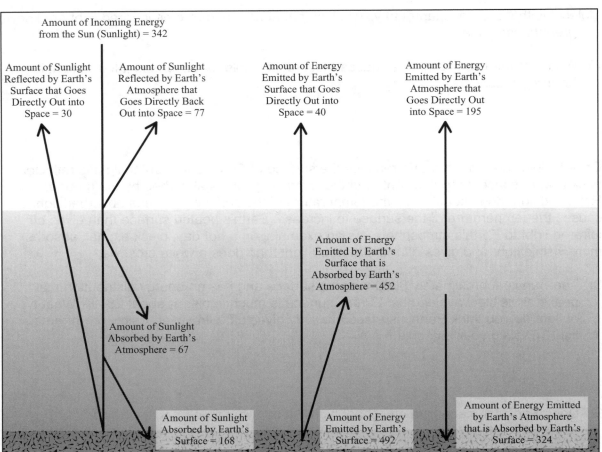

Amount of Incoming Energy
from the Sun (Sunlight) = 342

Amount of Sunlight
Reflected by Earth's
Surface that Goes
Directly Out into
Space = 30

Amount of Sunlight
Reflected by Earth's
Atmosphere that
Goes Directly Back
Out into Space = 77

Amount of Energy
Emitted by Earth's
Surface that Goes
Directly Out into
Space = 40

Amount of Energy
Emitted by Earth's
Atmosphere that
Goes Directly Out
into Space = 195

Amount of Energy
Emitted by Earth's
Surface that is
Absorbed by Earth's
Atmosphere = 452

Amount of Sunlight
Absorbed by Earth's
Atmosphere = 67

Amount of Sunlight
Absorbed by Earth's
Surface = 168

Amount of Energy
Emitted by Earth's
Surface = 492

Amount of Energy Emitted
by Earth's Atmosphere
that is Absorbed by Earth's
Surface = 324

Figure 3

Figure 3 shows how light/energy flows through the Earth system for the "greenhouse effect." The numbers listed describe the amount of energy flowing through the system (units of watts per square meter). A larger number indicates that more energy is flowing through that labeled pathway.

12) How does the total amount of energy coming from the Sun compare to the total amount of energy leaving Earth to space? Provide numbers to support your answer.

13) What type of light primarily heats Earth's surface and where does this light come from? What type of light primarily heats Earth's atmosphere and where does this light come from?

14) Is more energy absorbed by Earth's surface in the form of light coming from the Sun or from light emitted by Earth's atmosphere? Explain your reasoning, and provide numbers to justify your answer.

15) Due to the light absorbed by Earth's surface that was emitted by Earth's atmosphere, is Earth's temperature near the surface going to be warmer or cooler than it would be without this absorbed light?

16) Fill in the empty boxes in Figure 4 below with the correct type(s) of light. Use the abbreviations UV, IR, and VIS.

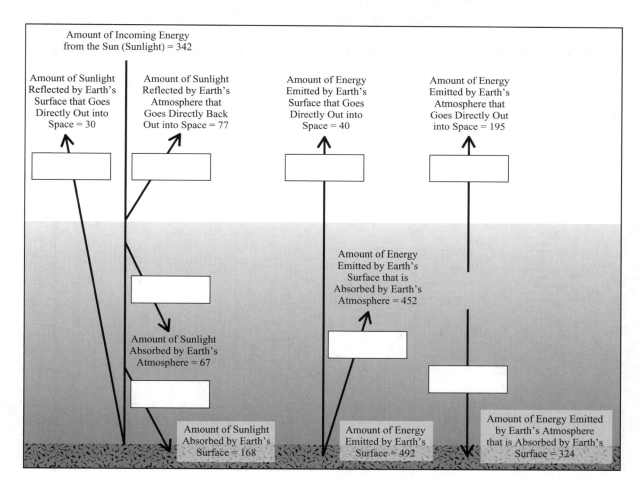

Figure 4

The flow of energy shown in Figures 3 and 4 is the source of the natural "atmospheric greenhouse effect." Visible light penetrates the atmosphere and is absorbed by the surface. The heated surface gives off infrared light that is then absorbed by the atmosphere. The heated atmosphere gives off infrared light out to space and also back down to Earth's surface, making the surface temperature warmer than it would be without a greenhouse effect. The amount of energy entering and leaving the Earth system can be balanced, but Earth's surface temperature is warmer because the surface is heated both by visible light from the Sun and infrared light sent back from the atmosphere.

17) Consider the following debate between two students regarding the greenhouse effect.

Student 1: *So the greenhouse effect is caused by infrared light being trapped in Earth's atmosphere. Visible light from the Sun heats the ground, but the infrared light given off by the ground gets permanently trapped in the atmosphere and can never escape.*

Student 2: *I think that's close. But based on Figure 3, all of the arrows balance and just as much energy leaves Earth as comes in. I think the greenhouse effect makes the surface hotter than it would be without greenhouse gases because the ground gets visible light from the Sun **AND** infrared light given off by the atmosphere that is sent back to the surface.*

Do you agree or disagree with either or both of the students? Explain your reasoning.

Consider the information provided in the graph and table below. The graph shows the temperature (expressed in kelvins) at different distances from the Sun (expressed in astronomical units or AUs) in the solar system during the time when the planets were originally forming. The table provides some common temperatures to use for comparison.

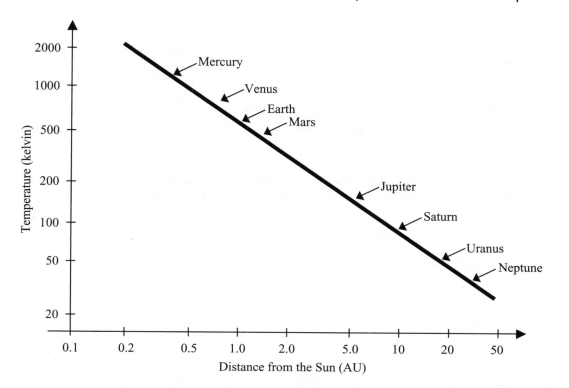

Condition	Temp. Fahrenheit	Temp. Celsius	Temp. kelvin
Severe cold temps. on Earth	−100	−73	199
Water freezes	32	0	273
Room temp.	72	22	296
Human body	98.6	37	310
Water boils	212	100	373

1) What was the temperature at the location of Earth?

2) What was the temperature at the location of Mars?

3) Which planets formed at temperatures hotter than the boiling point of water?

4) Which planets formed at temperatures cooler than the freezing point of water?

At temperatures hotter than the freezing point of water, light gases, such as hydrogen and helium, likely had too much energy to condense together to form the large, gas giant, Jovian planets.

5) Over what range of distances from the Sun would you expect to find light gases, such as hydrogen and helium, collecting together to form a Jovian planet? Explain your reasoning.

6) Over what range of distances from the Sun would you expect to find only solid, rocky material collecting together to form a terrestrial planet? Explain your reasoning.

7) Is it likely that a large, Jovian planet would have formed at the location of Mercury? Explain your reasoning.

Part I: Earth and Moon

Astronomers often deal with large numbers for distances, masses, and other quantities. They often use ratios to get a better sense of how big or small these quantities are. This can be useful in our daily lives as well. For example, we may not have a good sense for the length of a 40-meter-long commercial jet, but saying that the jet is approximately eight times longer than a car may be more meaningful to us. In this activity we will use ratios to try to better understand the size of objects in the solar system, in particular the size of the Sun.

Distances such as the following can be hard to conceptualize:

Moon's diameter: 3,476 km

Earth's diameter: 12,756 km

But we can think about these sizes in terms of one another so that we can create a scale model of Earth and the Moon in our minds. If we wish to express how many times bigger Earth is than the Moon, we can divide Earth's diameter by the Moon's diameter. The result is roughly 4 (12,756/3476 ≈ 4). This means Earth is approximately four times wider than the Moon, or equivalently, you could fit about four Moons across the diameter of Earth (as shown below).

1) Which of the following pairs of objects would make a good scale model of Earth and the Moon?

Earth

   a) a basketball and a soccer ball

   b) a basketball and a baseball (or softball)

   c) a basketball and a ping-pong ball

   d) a basketball and a pea

   e) a basketball and a grain of sand

The distance between Earth and the Moon is much larger than either the Moon or Earth—but how much larger? If we divide the distance between Earth and the Moon (384,000 km) by Earth's diameter, we get 384,400/12,756 ≈ 30. This means you could fit approximately 30 Earths in the space between Earth and the Moon.

2) Using small circles to represent Earth and the Moon, sketch a **scale model** of the Earth–Moon orbital system below. Be sure your scale model correctly shows the two scale ratios described above.

3) To make a scale model of the Earth–Moon orbital system, you not only need to pick appropriately sized objects to represent Earth and the Moon, you also need to place them the correctly scaled distance apart. Let's say you use a 1-foot (12-inch) basketball and a 3-inch orange as your Earth and Moon, respectively. About how far apart must they be placed to represent an accurate scale model of the Earth–Moon orbital system? (*Circle your answer below.*) Explain your reasoning.

   a) 1 foot

   b) 4 feet

   c) 10 feet

   d) 30 feet

   e) 300 feet

Part II: The Sun

Compared to the size of Earth, the Sun (with a diameter 1,392,000 km) is about 110 times bigger than Earth ($1,392,000/12,756 \approx 110$).

4) Can any combinations of the following items be used to make an accurate scale model of Earth and the Sun? If so, which two would you choose and why? If not, why not?

   a) basketball

   b) soccer ball

   c) baseball (or softball)

   d) ping-pong ball

   e) pea

   f) grain of sand

Now let's compare the Sun's diameter to the size of the Moon's orbit around Earth. The diameter of the Moon's orbit around Earth is about 769,000 km across. So, the ratio of the Sun's diameter to the Moon's orbital diameter is roughly 2 ($1,392,000/769,000 \approx 2$).

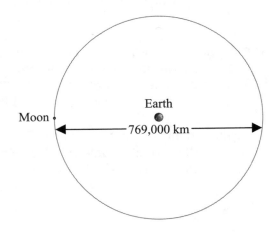

5) Does this mean that two Suns placed side-by-side would fit inside the Moon's orbit around Earth, or that two Moon orbits placed side-by-side would fit across the Sun? Draw a sketch below to illustrate your answer.

The distance from Earth to the Sun averages about 150,000,000 km. This makes the distance between the Sun and Earth about 110 times larger than the diameter of the Sun (150,000,000/1,392,000 ≈ 110).

6) If you were to use a 1-foot (12-inch) basketball to represent the Sun, how far would it have to be from Earth to be an accurate scale model?

    a) 1 foot

    b) 10 feet

    c) 30 feet

    d) 110 feet

    e) 300 feet

7) If we used a basketball to represent the Sun and a ping-pong ball to represent Earth, and separated them by the distance you answered in Question 6, would we have an accurate scale model of the Sun–Earth system? Explain your answer.

8) How many Moons would fit across the diameter of the Sun?

9) Approximately how many times could the Moon's orbital diameter fit between Earth and the Sun?

Use the H–R diagram below to answer questions throughout this activity.

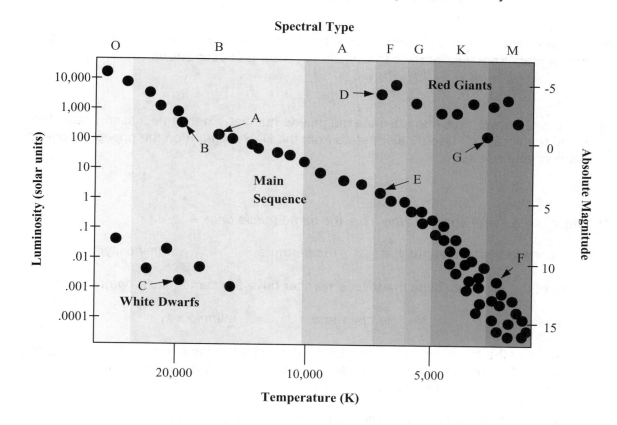

1) What are the spectral type, temperature, absolute magnitude number, and luminosity of Star A?

a) Spectral type:

b) Temperature:

c) Absolute magnitude:

d) Luminosity:

2) Which two pairs of labeled stars (A–G) in the diagram have the same temperature?

3) Do stars of the same temperature have the same spectral type? Use a pair of stars from your answer to Question 2 to support your answer.

4) Which two pairs of labeled stars have the same luminosity?

5) Do stars with the same luminosity have the same absolute magnitude number? Use a pair of stars from your answer to Question 4 to support your answer.

6) If two stars have the same absolute magnitude number, do they necessarily have the same temperature? Use a pair of stars from the H–R diagram on the previous page to support your answer.

7) Stars of the same spectral type have the same (*circle one*):

    absolute magnitude number   temperature   luminosity

8) Stars of the same absolute magnitude number have the same (*circle one*):

    spectral type   temperature   luminosity

9) For each of the following star descriptions, state whether the star would be a red giant, white dwarf, or main sequence star, and provide the letter(s) of a star from the H–R diagram that fits each description.

    a) very bright (high luminosity) and very hot (high temperature)

    b) very dim and cool

    c) very dim and very hot

    d) very bright and cool

Stars begin life as a cloud of gas and dust. The birth of a star begins when a disturbance, such as the shock wave from a supernova, triggers the cloud of gas and dust to collapse inward.

1) Imagine that you are observing the region of space where a cloud of gas and dust is beginning to collapse inward to form a star (the object that initially forms in this process is called a protostar). Will the atoms in the collapsing cloud move away from one another, move closer to one another, or stay at the same locations?

2) What physical interaction, or force, causes the atoms to behave as you described they would in Question 1?

3) Would you expect the temperature at the center of the protostar to increase or decrease with time? Explain your reasoning.

The inward collapse of material causes the center of the protostar to become very hot and dense. Once the central temperature and density reach critical levels, nuclear fusion begins. During the fusion reaction, hydrogen atoms are combined together to form helium atoms. When this happens, photons of light are emitted. Once the outward pressure created by the energy given off during nuclear fusion balances the inward gravitational collapse of material, a state of *hydrostatic equilibrium* is reached, and the star no longer collapses. When this happens, the protostar becomes a main sequence star.

Consider the information shown in the table below when answering Questions 4 through 7.

Mass of the Star (in multiples of Sun masses, M_{sun})	Approximate Main Sequence Lifetime of the Star
0.5 M_{sun}	50 billion years
1.0 M_{sun}	10 billion years
2.0 M_{sun}	2 billion years
6.0 M_{sun}	110 million years
60 M_{sun}	360 thousand years

4) Which live longer, high-mass or low-mass stars?

5) Based on your answer to Question 4, do you think that the rate of nuclear fusion in a high-mass star is greater than, less than, or equal to the rate of nuclear fusion in a low-mass star?

6) Which of the following statements best describes how the lifetimes compare between Star A (a star with a mass equal to the Sun) and Star B (a star with six times the mass of the Sun)? Circle the best possible response given below. (Note: It may be helpful to examine the information given in the table on the previous page.)

   a) Star A will live less than 1/6th as long as Star B.

   b) Star A will live 1/6th as long as Star B.

   c) Star A will have the same lifetime as Star B.

   d) Star A will live six times longer than Star B.

   e) Star A will live more than six times longer than Star B.

   Explain your reasoning for the choice you made.

7) The Sun has a lifetime of approximately 10 billion years. If you could determine the rate of nuclear fusion for a star with twice the mass of the Sun, which of the following would best describe how its fusion rate would compare to the Sun? Circle the best possible response to complete the sentence given below. (Note: It may be helpful to examine the information given in the table on the previous page.)

   A star with twice the mass of the Sun would have a rate of nuclear fusion that is _____ the rate of fusion in the Sun.

   a) less than

   b) a little more than

   c) twice

   d) more than twice

   Explain your reasoning for the choice you made.

1) Imagine we measured the light emitted by a Sun-like (G-spectral type) main sequence star for several weeks. Which of the graphs below most likely shows how its graph of brightness versus time would look (*circle A or B*)?

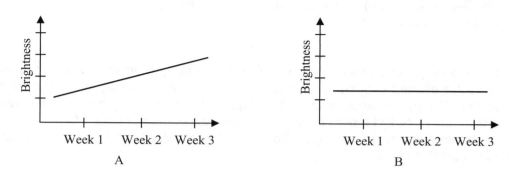

2) Imagine instead that we measure the light emitted by an A-spectral type main sequence star at the same distance as the Sun-like (G-spectral type) star from Question 1 for several weeks. Compared to the graph of the Sun-like star you chose above, which of the graphs below most likely shows how the graph of brightness versus time would look for an A-spectral type star (*circle C or D*)?

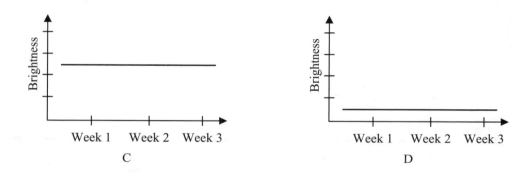

3) Imagine that these two stars are actually quite close together such that the total amount of light received from the pair can be shown in a single graph. Compared to the graphs you selected in Questions 1 and 2, which of the graphs below most likely shows how the combined graph of brightness versus time for the two stars would look (*circle E or F*)?

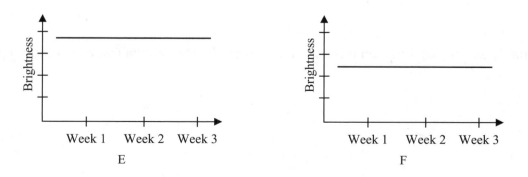

4) Imagine that the objects shown at the right represent the Sun-like and A-spectral type stars from the previous questions. Label which object would best represent the Sun-like (G-spectral type) star and which object would best represent the A-spectral type star.

5) Stars that are very close together will often orbit around one another, and, occasionally, their orbits are aligned in such a way that one star will pass directly in front of the other as seen from Earth. These stars are often referred to as eclipsing binary stars. Which of the two times (1 or 2) labeled below most likely indicates the time when the Sun-like (G-spectral type) star was passing **in front** of the A-spectral type star from the previous questions. (*circle 1 or 2*)?

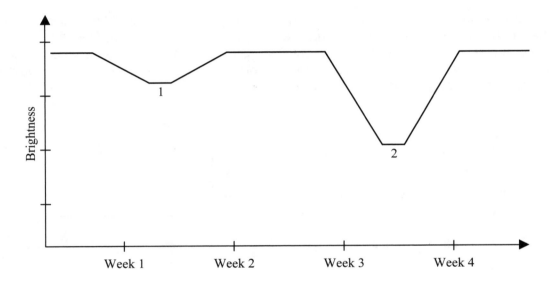

Explain your reasoning.

6) What is the physical reason the bottom of the dip is a horizontal line rather than a point?

7) Imagine that you are watching a binary star system containing an M-spectral type main sequence star and a B-spectral type main sequence star as they each complete one full orbit. During this time, you are able to see the stars entirely separate from one another. At another time, you see the B-spectral type star in front of the M-spectral type star. And at an entirely different time, you see that the B-spectral type star has moved behind the M-spectral type star. In the space below, draw three sketches showing what the stars would look like at the three times described.

8) At which of the times you drew would you measure the greatest amount of light coming to you? Explain your reasoning.

9) At which of the times you drew would you measure the least amount of light coming to you? Explain your reasoning.

10) Two students are talking about how the light curve would appear when observing the eclipsing binary star system described in Question 7.

Student 1: *I think the dip in the graph is deepest when the blue star passes in front of the red star. Since the blue star is so much bigger, it will block off all of the light from the red star.*

Student 2: *I disagree, a hot star emits way more light from each part of its surface than a cold star does. So I think the deepest dip will happen when the cold star blocks the light from the hot star.*

Do you agree or disagree with either or both of the students? Explain your reasoning.

Binary Stars

11) In the graph below, draw a line to illustrate the amount of light an Earth observer would detect from an eclipsing binary star system that contains a Sun-like (G-spectral type) star and a red giant (with an orbital period of four weeks). Clearly label the dips as to which star is in front and which is behind.

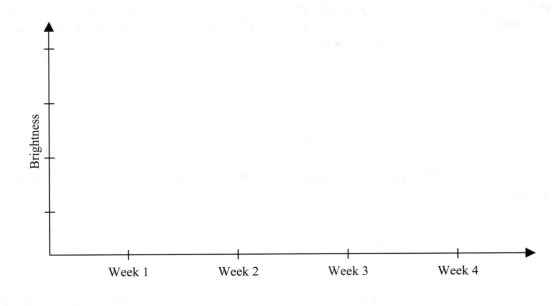

Explain your reasoning.

Part I: Extrasolar Planet Systems' Properties of Motion & Doppler Shift

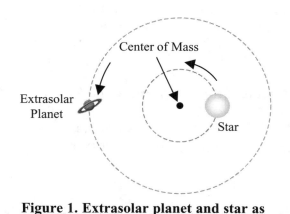

Figure 1. Extrasolar planet and star as seen from above. (not to scale)

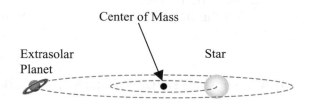

Figure 2. Extrasolar planet and star as seen edge-on or from the side. Note that the extrasolar planet is moving toward you.
(not to scale)

Figure 1 and Figure 2 show the orbits of the same extrasolar planet and star from two different points of view.

1) As an extrasolar planet orbits around a star, the gravitational attraction between the two objects causes the star to make a small orbit around the system's center of mass. Which object travels in the largest orbit (*circle one*)?

the extrasolar plant	the star	they both have the same size orbit	you can't determine which has the larger orbit

2) Which object takes a greater amount of time to complete one orbit (*circle one*)?

the extrasolar planet	the star	they both take the same amount of time	you can't determine which takes longer

Explain your reasoning.

3) At the instant shown in Figure 1, which direction is the extrasolar planet moving (*circle one*)?

toward the bottom   toward the top   toward the
    of the page   of the page   star

4) At the instant shown in Figure 1, which direction is the star moving (*circle one*)?

toward the bottom   toward the top   toward the
    of the page   of the page   extrasolar planet

5) In general, how does the direction the extrasolar planet is moving compare with the direction the star is moving?

6) Figure 2 shows the extrasolar planet and star from the side or as seen edge-on. At the instant shown, which direction is the planet moving (*circle one*)?

coming out of the page   moving into the page directly   toward the central star
   directly toward you   away from you

7) Two students are having a discussion about the relationship between the movement of the central star and extrasolar planet and the Doppler shift of the light coming from the star.

     Student 1:   *Since Figure 2 states that the extrasolar planet is moving out of the page, directly toward us, then the light from the star we observe will be blueshifted.*
     Student 2:   *I disagree, the light from the star will be redshifted because the star is moving in the opposite direction the planet is moving.*

Do you agree or disagree with either or both of the students? Explain your reasoning.

8) Would the light from the star in Figure 1 be blueshifted, redshifted, or not shifted? Explain your reasoning.

9) Would the light from the star in Figure 2 be blueshifted, redshifted, or not shifted? Explain your reasoning.

10) If you are unable to detect any Doppler shift from a star in an extrasolar planet system, how must this system be oriented with respect to your line of sight? Explain your reasoning and include a drawing to illustrate your answer in the space below.

Part II: Evaluating Extrasolar Planet Systems

The amount that the light from a star in an extrasolar planet system will be Doppler shifted depends on the mass of the star M_{star}, the mass of the planet m_{planet}, and the distance d between the star and the planet. This relationship can be written as:

$$\text{Amount of Doppler shift in stars's light} \propto \frac{m_{planet}}{\sqrt{M_{star}}\,d}$$

Figure 3 below shows four different extrasolar planet systems (A–D). Use this figure to answer Questions 11–16.

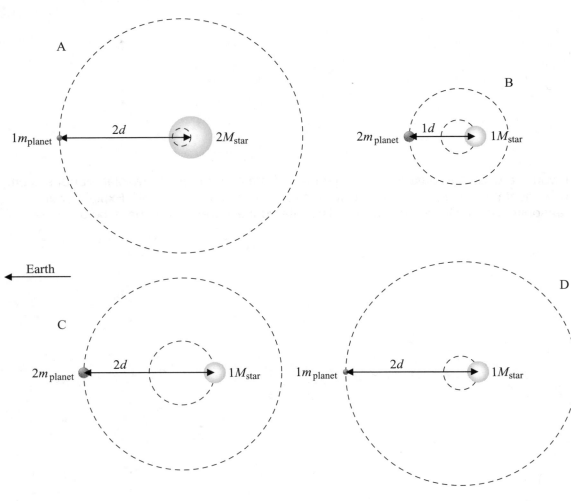

Figure 3

11) Which extrasolar planet system(s) (A–D) has the lowest mass star?

12) Which extrasolar planet system(s) (A–D) has the highest mass planet?

13) In which extrasolar planet system(s) (A–D) is the planet closest to the star?

14) In which extrasolar planet system(s) (A–D) would we receive light from the star with the largest Doppler shift? Explain your reasoning.

15) Which system (A–D) has the extrasolar planet that is easiest to detect from Earth? Explain your reasoning.

16) Two students are discussing their answers to Questions 14 and 15.

Student 1: *I think Extrasolar Planet System C shows the star with the largest Doppler shift. This is because the star in System C has the largest orbit. This means that this extrasolar planet will be the easiest to detect.*

Student 2: *I don't think that Doppler shift is caused by the size of the star's orbit. To cause a large Doppler shift, you want a low-mass star that is close to a large-mass planet, and that is Extrasolar Planet System B.*

Do you agree or disagree with either or both of the students? Explain your reasoning.

Figure 4 below shows the graph of four radial velocity curves (E–H) for the four stars in Figure 3. Use these graphs to answer Questions 17 and 18.

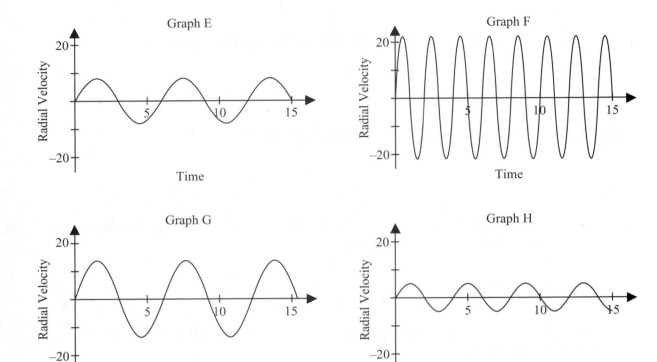

Figure 4

17) Match each graph (E–H) with the extrasolar planet systems (A–D) from Figure 3. Explain your reasoning.

Extrasolar Planet System A:

Extrasolar Planet System B:

Extrasolar Planet System C:

Extrasolar Planet System D:

LECTURE-TUTORIALS FOR INTRODUCTORY ASTRONOMY
THIRD EDITION

18) On the graph (E, F, G, or H) that depicts the largest Doppler shift:

   a) draw a circle on the curve at each time that corresponds with the **star** moving with its fastest speed toward Earth. Explain your reasoning.

   b) draw a triangle on the curve at each time that corresponds with the **extrasolar planet** moving with its fastest speed toward Earth. Explain your reasoning.

Use Figure 5 to answer the following question.

19) Given the location marked with the dot on the star's radial velocity curve, at which location (I–L) would you expect the planet to be located at this time? Explain your reasoning.

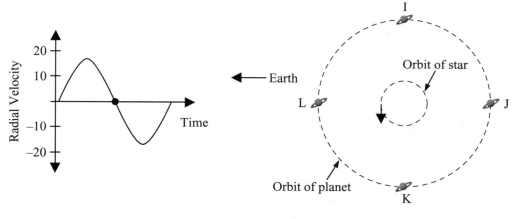

Figure 5

Main sequence stars that can no longer support nuclear fusion of hydrogen in their cores will become red giant stars. Although most main sequence stars become red giants, their specific evolutionary paths after this red giant phase vary greatly, depending on mass.

A low-mass star, less than about eight times the mass of our Sun ($<8m_s$), eventually ejects its outer layers to produce a planetary nebula. The stellar core remaining in the middle of this planetary nebula is called a white dwarf.

In contrast, a high-mass star, more than eight times the mass of the Sun ($>8m_s$), will eventually explode as a Type II supernova. Depending on the original mass of the star, the Type II supernova will leave behind either a neutron star or, if the original star was extremely massive, a black hole.

1) Use the information above and the word list below to fill in the ovals in the diagram on the next page. Be sure to look at the arrows and words between the ovals to make sure these links between ovals make sense. Check your work with another group.

Word list:
neutron star
black hole
planetary nebula
white dwarf
nova
Type II supernova

The diagram does not give us all the information known about the death of stars. Since it is incomplete, we can always add to this diagram as we learn more information.

2) In Parts a and b below, you are given some additional information about the end states of stars. Your task is to change or add to the diagram to incorporate this additional information. (Note: There are several ways to accomplish this.)

a) If a white dwarf has a nearby binary companion star, it can gravitationally attract material from its companion in a process known as accretion. When the white dwarf accretes enough material from the companion, the white dwarf will either (1) blow off the outer layers of accreted material in a controlled fusion reaction known as a nova, leaving behind the white dwarf unchanged, or (2) experience a violent, uncontrolled fusion reaction that causes the white dwarf and accreted material to explode as a Type Ia supernova, destroying the white dwarf and leaving nothing behind.

b) Completely by itself in space, a black hole can be nearly impossible to detect. However, if a black hole has a nearby binary companion star, the strong gravitational pull of the black hole can gravitationally attract material from its companion in a process known as accretion. This material then spirals around the black hole and increases in temperature. This process causes the rapidly moving material to emit large amounts of X-ray radiation, which we can detect with X-ray telescopes. Thus, one way to look for black holes is to look for strong X-ray sources.

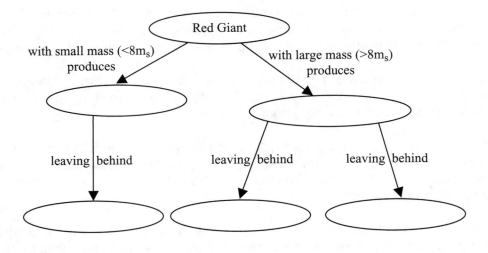

This tutorial will give you a better understanding of the size of the Milky Way Galaxy by investigating the distances and sizes of objects within the Milky Way Galaxy and outside the Milky Way Galaxy elsewhere in the universe. Because we are located within the Milky Way, we are unable to take a picture of our entire galaxy from the outside. Below is a picture of a spiral galaxy similar to the Milky Way. Let's assume that this picture represents our Milky Way Galaxy and has the dimensions labeled below. **Note that in this picture, 1 centimeter (cm) represents 10,000 light-years (ly); equivalently, you can use 1 millimeter (mm) to represent 1,000 light-years (ly).**

1) The Sun's position in the Milky Way is shown in the picture above. What is the approximate distance from the Sun to the center of the Milky Way? Recall that 1 cm represents 10,000 ly.

2) The table below lists five bright stars in the night sky. Write the letter of the dot (A–E) from the picture on the previous page that best represents the location of each star. You can use letters more than once. Recall that 1 mm represents 1,000 ly.

Star	Distance from Sun (in light-years)	Letter
Sirius	9	
Vega	26	
Spica	260	
Rigel	810	
Deneb	1,400	

3) We normally consider Deneb to be a bright but distant star at 1,400 ly away from the Sun. Compared to the size of our Milky Way Galaxy, is Deneb truly distant? Explain your reasoning.

4) Are the stars from Question 2 inside or outside the Milky Way Galaxy? Explain your reasoning.

5) The table below lists three Messier objects and their distances from the Sun. Write the letter of the dot (A–E) from the picture on the previous page that best represents the location of each object. You can use letters more than once.

Messier Object	Distance from Sun (in light-years)	Letter
M45 Open Cluster (Pleiades)	380	
M1 (Crab Nebula)	6,300	
M71 Globular Cluster	12,700	

6) Are these Messier objects part of the Milky Way Galaxy? Explain your reasoning.

7) The Crab Nebula has a width of about 11 light-years. If you wanted to accurately draw the Crab Nebula on your diagram, would you use a small blob or a tiny dot at the location you indicated in Question 5? Explain your reasoning. Note: The dots marking the locations on the picture are about 1 mm across.

8) The Sun is much smaller than a nebula. We used a dot to represent the Sun's location in the picture. Is this dot too small, too large, or just the right size to represent the size of the Sun in the picture? Explain your reasoning.

9) The Milky Way Galaxy is one of the largest galaxies in a group of nearby galaxies called the Local Group. The following table lists the distances to the centers of three Local Group galaxies. Draw a dot on your picture (if possible) to represent the center of each galaxy. Don't worry about the direction (left/right/up/down) for each galaxy; just place a dot an appropriate distance from the Sun.

Galaxy	Distance from Sun (in light-years)
Sagittarius Dwarf Elliptical Galaxy (SagDEG)—closest galaxy to Milky Way	80,000
Large Magellanic Cloud (LMC)	160,000
Andromeda Galaxy (M31)	2,500,000

Do any of these galaxies fit on the page? Which one(s)?

10) Are the objects listed in Question 9 inside or outside the Milky Way? Explain your reasoning.

11) SagDEG is approximately 11,000 ly across. Is this galaxy better represented on your diagram by a small blob or a tiny dot? Explain your reasoning, and make an appropriate sketch to represent the galaxy.

12) Within the Local Group, the two largest galaxies are the Milky Way and Andromeda Galaxies. From Question 9, we saw that the Andromeda Galaxy was about 2,500,000 ly from us. On the picture, this location would be 250 cm (about two-and-a-half meters or 8 feet) away from the dot representing the Sun.

The nearest group of galaxies to us (not counting our own Local Group) is the Virgo Cluster, about 60,000,000 ly away. How many centimeters away would this cluster be on our picture? How many meters away would this be?

When stars are viewed through a telescope, they typically appear as bright points of light without any apparent size or structure. However, there are some objects in the sky that, when viewed through a telescope, look "fuzzy" and cloudlike. Some of these objects, like those shown in the *Hubble Space Telescope* image to the right, are actually galaxies (containing billions of stars) that are much farther from us than the individual stars we see in the night sky.

Part I: Applying Hubble's Classification Scheme

1) Using the images of galaxies provided on the inside back cover of your *Lecture-Tutorial* book, sort these galaxies (using Hubble's categories) as being either an elliptical or a spiral galaxy. Use the table below to record your results. Try to find patterns in terms of shape, size, color, or any other distinct features that help in sorting the galaxies.

Hubble's Categories	Galaxy ID Numbers	Defining Characteristics (Describe the characteristics that you used to distinguish one class of galaxy from the other)
Elliptical		
Spiral		

Part II: Understanding the Types of Galaxies

In Part I you classified the galaxies into different categories according to their appearance, or *morphology*. We will now investigate what a galaxy's morphology can tell us about its physical characteristics. These physical characteristics include: (a) the ages of the stars in the galaxy; (b) the presence or absence of dust in the galaxy; and (c) the presence or absence of gas and star formation. Keep in mind that these properties are linked together in a physical way. The objective of this activity is for you to learn how these characteristics are related to galaxy classification and morphology.

The Ages of Stars:

2) Which of the galaxies appear to be mostly red? (Note: The word "red" is used to also include the colors orange and yellow.) Record the number and classification (elliptical or spiral) of each galaxy. Why do you think these galaxies appear red?

3) Which of the galaxies appear to be mostly blue? Record the number and classification (elliptical or spiral) of each galaxy. Why do you think these galaxies appear blue?

4) Which types of galaxies appear to have many young stars: elliptical, spiral, or both? Explain your reasoning.

5) Do the galaxies that you identified in Question 4 also contain old stars? Explain your reasoning.

Dust in Galaxies:
Besides stars, galaxies sometimes also contain dust. This dust produces dark bands across, or patches in, the galaxy.

6) Which of the galaxies show evidence of dust? Record the number and classification (elliptical or spiral) of each galaxy. Explain your reasoning.

Gas and Star Formation in Galaxies:
In addition to stars and dust, galaxies may also contain gas.

7) Would you say that a galaxy that is experiencing active star formation contains little or abundant gas? Explain your reasoning.

8) Which type of galaxy (elliptical or spiral) would have abundant gas available? Explain your reasoning.

9) Which type of galaxy (elliptical, spiral, both, or neither) is likely to contain both O-spectral type stars as well as M-spectral type stars? Explain your reasoning.

10) Which type of galaxy (elliptical, spiral, both, or neither) is likely to contain many M-spectral type stars but very few (if any) O-spectral type stars? Explain your reasoning.

11) Which type of galaxy (elliptical, spiral, both, or neither) is likely to contain only O-spectral type stars? Explain your reasoning.

12) Consider the discussion among three students about a galaxy that appears red.

Student 1: *Because there is mainly red light in this galaxy and no blue light, I think that only small, red stars formed in this galaxy and not any big, blue ones.*

Student 2: *I disagree. It's just that blue stars don't last very long. I think the blue stars that may have been there in the past have already evolved into red giants, so the galaxy looks red due to the light from all the red giants.*

Student 3: *Wait a minute. I think you are both wrong. I thought that both blue stars and red giants live short lives, so they should both be gone. I think that all the blue stars that formed early on have evolved into the red stars that are there now. So the galaxy appears red because it's full of a lot of old, red stars that used to be the blue stars.*

Do you agree or disagree with any or all of the students? Explain your reasoning.

13) Hubble imagined the tuning fork diagram (shown at right) as representing an evolutionary sequence for galaxies, with galaxies starting off as elliptical and developing more structure over time. Do you think Hubble's proposed evolutionary sequence is correct? Why or why not?

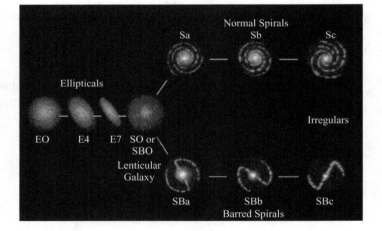

Part I: Motions of Planets

An object's orbit depends on the "mass inside" its orbit (also known as the *interior mass*). For a planet in our solar system, you can find the interior mass by adding the Sun's mass to the mass of each object between the Sun and the planet's orbit. For example, the interior mass to Earth's orbit would be the Sun's mass plus the mass of Mercury plus the mass of Venus.

Here is a table that lists each planet, the mass inside each planet's orbit, and the speed at which the planets orbit the Sun.

Planet	Interior Mass (solar masses)	Orbital Speed (km/s)
Mercury	1.00	47.9
Venus	1.00000017	35.0
Earth	1.0000026	29.8
Mars	1.0000056	24.1
Jupiter	1.0000059	13.1
Saturn	1.00096	9.66
Uranus	1.0012	6.81
Neptune	1.0013	5.43

1) Where is the vast majority of mass in the solar system located? What object or objects account for most of this mass?

2) Two students are discussing their answers to Question 1:

Student 1: *I think the majority of the mass in the solar system must include both the Sun and the planets. As you get farther away from the Sun, the interior mass gets bigger and bigger because you include more planets.*

Student 2: *I disagree. The majority of the mass in the solar system is from just the Sun by itself. Sure the mass gets a little bigger as you include more planets, but the additional mass from planets is really small.*

Do you agree or disagree with either or both of the students? Explain your reasoning.

3) How do the orbital speeds of planets farther from the Sun compare to the orbital speeds of planets closer to the Sun?

A planet's orbital speed depends on the gravitational force it feels. The strength of the gravitational force depends on the amount of mass that is inside of the planet's orbit as well as how far away the planet is from this interior mass.

4) How does the gravitational force on a planet far from the Sun compare to the gravitational force on a planet close to the Sun? Explain your reasoning.

5) Complete the blanks in the sentences of the following paragraph by either writing in the necessary information or circling the correct response in the parentheses (). It may be helpful to base your responses on the information provided in the previous table and your answers to the previous questions.

There are _____ planets inside Neptune's orbit and _____ planets inside Mercury's orbit. However, the interior mass for Neptune is _____ (much greater than/approximately the same as/much less than) the interior mass of Mercury. Neptune is _____ (much closer to/ much farther from/about the same distance from) the Sun as/than Mercury. Therefore the gravitational force exerted on Neptune is _____ (stronger/weaker/about the same strength) as/than the force exerted on Mercury. As a result, Neptune has an orbital speed that is _____ (much slower/much faster/about the same speed) as/than the orbital speed of Mercury.

Imagine you were able to add a very, very large amount of mass distributed evenly *between* the orbits of Jupiter and Saturn.

6) Which planet(s) will experience an increase in gravitational force and an increase in orbital speed from this added mass? Explain your reasoning.

Part II: Motions of Stars

One way to estimate the amount of mass in a spiral galaxy is by looking at how much light the galaxy emits. Where there is more light, there must be more stars and hence more mass. When astronomers measure the amount of light emitted by different regions of a galaxy, they often find that more light is emitted from the center of the galaxy and less light is emitted from the outer regions.

7) Based on the information provided above, where do you expect most of the mass of a galaxy to be located?

At right is a picture of a spiral galaxy similar to the Milky Way. The orbits of three stars are labeled. Star A is a star on the edge of the Milky Way's bulge. The Sun's orbit is shown at approximately the correct position. Star B is a star located farther out in the disk than the Sun.

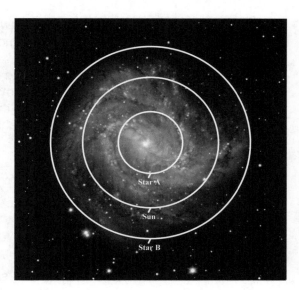

8) Based on your answer to Question 7 and the location of each star from the center of the galaxy, rank how you think the orbital speeds of Star A, Star B, and the Sun would compare from fastest to slowest. Explain your reasoning.

A graph of the orbital speed of stars versus their distance from the galaxy's center is called a *rotation curve*. Here are two possible rotation curves.

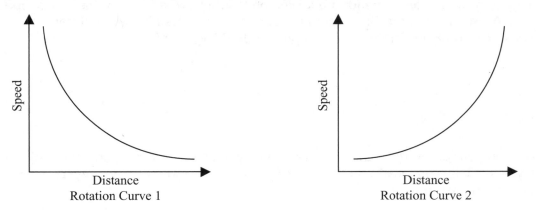

9) Which of the above rotation curves best represents the relationship you described in Question 8? Explain your reasoning.

Astronomers were surprised when they saw the real rotation curve for the Milky Way Galaxy (MWG). The rotation curve at the right is more like the MWG's real rotation curve.

10) Based on their orbital distance from the center of the galaxy, make three dots on the above rotation curve to represent Star A, Star B, and the Sun. Be sure to label which mark belongs to each star.

Edge of central bulge

Speed

Distance

11) Describe how the stars' orbital speeds shown in the real rotation curve for the MWG are different from the orbital speeds shown in the rotation curve you chose in Question 9.

12) Using the real rotation curve for the MWG, provide a new ranking for the orbital speeds of Star A, Star B, and the Sun, from fastest to slowest. Describe any differences between this ranking and the one you provided in Question 8.

13) Based on your answers to Question 12, would you say that most of the mass of the Milky Way Galaxy is located at its center (as is the case with our solar system)? Explain your reasoning.

14) Based on the MWG's real rotation curve and your answers to Questions 11–13, is the gravitational force felt by the MWG's stars greater than, less than, or about the same as what you expected from Questions 8 and 9? Explain your reasoning.

15) Two students are debating their answers to the previous questions:

Student 1: *Stars far from the center of the Milky Way are all moving at about the same speed. If most of the Milky Way's mass is located in its center, then stars far away from the center would orbit slower than stars closer to the center. Since this is not what we see, this must mean there is more mass throughout the outer regions of the galaxy than we can see. This also means that the Milky Way's stars feel a greater gravitational force than we originally expected.*

Student 2: *I disagree. There are fewer stars in the outskirts of the Milky Way than in the center, so there's less mass out there than at the center. Most of the Milky Way's mass must be at its center. So, since the stars are all going about the same speed, where the mass is located must not affect their speed. The gravitational force these stars feel probably gets weaker just like we would expect.*

Do you agree or disagree with either or both of the students? Explain your reasoning.

16) Astronomers initially thought there was more mass in the center of the galaxy than in the rest of the galaxy because there's more light coming from the center. Describe how astronomers' observations show these initial ideas are wrong.

17) Is there more or less mass in the Milky Way's disk and halo than we can see? Explain your reasoning.

Imagine that you have received six pictures of six different children who live near six of the closest stars to the Sun. Each picture shows a child on his or her 12th birthday. The pictures were each broadcast directly to you (using a satellite) on the day of the child's birthday. Note the abbreviation "ly" is used below to represent a light-year.

- Melissa lives on a planet orbiting Ross 154, which is 9.5 ly from the Sun.
- Max lives on a planet orbiting Barnard's Star, which is 6.0 ly from the Sun.
- Jade lives on a planet orbiting Sirius, which is 8.6 ly from the Sun.
- Sydney lives on a planet orbiting Alpha Centauri, which is 4.3 ly from the Sun.
- Joyce lives on a planet orbiting Epsilon Eridani, which is 10.8 ly from the Sun.
- Crystal lives on a planet orbiting Procyon, which is 11.4 ly from the Sun.

1) Describe in detail what a light-year is. Is it an interval of time, a measure of length, or an indication of speed? It can only be one of these quantities.

2) Which child lives closest to the Sun? How far away does he or she live?

3) What was the greatest amount of time that it took for any one of the pictures to travel from the child to you?

4) If each child was 12 years old when he or she sent his or her picture to you, how old was each of the children when you received their picture?

Melissa _____   Max _____   Jade _____
Sydney _____   Joyce _____   Crystal _____

5) Is there a relationship between the actual age of each child when you received their picture and his or her distance away from Earth? If so, describe this relationship.

6) Imagine that the six pictures were broadcast by satellite to you and that they all arrived at exactly the same time. For this to be true, does that mean that all of the children sent their pictures at the same time? If not, which child sent his or her picture first and which child sent his or her picture last?

LECTURE-TUTORIALS FOR INTRODUCTORY ASTRONOMY
                                                              THIRD EDITION

7) The telescope image at the right was taken of the Andromeda Galaxy, which is located about 2.5 million ly away from us. Is this an image showing how the Andromeda Galaxy looks right now, how it looked in the past, or how it will look in the future? Explain your reasoning.

8) Imagine that you are observing the light from a distant star that was located in a galaxy 100 million ly away from you. By analysis of the starlight received, you are able to tell that the image we see is of a 10-million-year-old star. You are also able to predict that the star will have a total lifetime of 50 million years, at which point it will end in a catastrophic supernova.

    a) How old does the star appear to be to us here on Earth?

    b) How long will it be before we receive the light from the supernova event?

    c) Has the supernova already occurred? If so, when did it occur?

9) Imagine that you take images of two main sequence stars that have the same mass. From your observations, both stars *appear* to be the same age. Consider the following possible interpretations that could be made from your observations.

    a) Both stars are the same age and the same distance from you.

    b) Both stars are the same age but at different distances from you.

    c) The stars are actually different ages but at the same distance from you.

    d) The star that is closer to you is actually the older of the two stars.

    e) The star that is farther from you is actually the older of the two stars.

    How many of the five choices (a–e) are possible? Which ones? Explain your reasoning.

Part I: The Observable Universe

Each tiny dot in the picture below represents a galaxy. The Milky Way Galaxy is represented by a tiny dot at the center of the picture. All of the galaxies inside the circle can be seen from Earth. The circumference of this circle defines what is called our *observable universe*. Any galaxy that exists outside the circle is so far away that its light has not had time to reach Earth and is therefore not part of our observable universe.

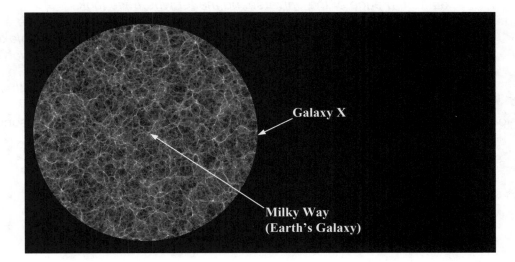

Galaxy X

Milky Way
(Earth's Galaxy)

1) Do you think the galaxies we can see from Earth are the only galaxies in the *entire universe*? Explain your reasoning.

2) Draw a circle around Galaxy X that represents its *observable universe*.

3) Is the *observable universe* that you drew for Galaxy X different in size than the *observable universe* for Earth? Explain your reasoning.

4) Two students are talking about the *observable universe* for Galaxy X:

Student 1: *People living in Galaxy X have a strange view of the universe. When they look in one direction, they see a bunch of galaxies, but when they look in the other direction, all they see is empty space. Galaxy X must be at the edge of the universe since there's nothing but black, empty space beyond it. We're lucky we live at the center since we can see galaxies all the way out to the edge of the universe, no matter where we look.*

Student 2: *I think you're wrong. People living in Galaxy X would probably see a bunch of galaxies in every direction they look, but they can see some galaxies that we can't, just like we can see galaxies they can't. The observable universe for any galaxy should look similar to ours. I don't think we are at the center of the universe and I don't think Galaxy X is at the edge either.*

Do you agree or disagree with either or both of the students? Explain your reasoning.

Part II: An Analogy for Expansion

One way to try to understand and envision the expansion of the universe is by creating analogies that model the different aspects of our real expanding universe. One way to model the expanding universe is to use a "balloon" analogy. In this analogy, the space and time of the universe are modeled by the "surface" or "skin" of an expanding balloon. The *entire universe* exists only on the surface of the balloon. Light can travel only on the surface of the balloon.

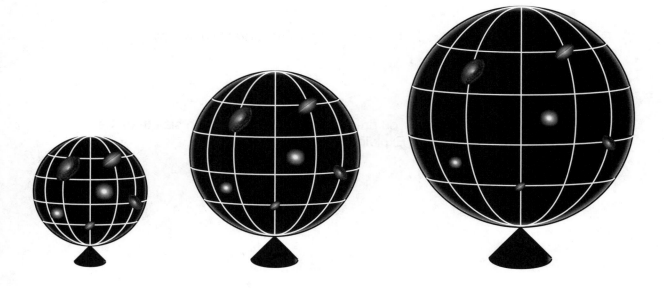

5) Do objects, light, or events in the *entire universe* also exist inside or outside of the balloon's surface in this analogy?

6) If you were to travel from galaxy to galaxy along the surface of the balloon, would you ever encounter an edge?

7) If you were to travel over the entire surface of the balloon universe, would you ever find a location that is the center of the *entire universe*?

8) Consider the following debate between two students about their answers to the previous questions:

Student 1: *Someone living on the surface of this balloon universe will definitely encounter an edge and a center. All they have to do is look from their location across the inside of the balloon to a location on the other side. The center of the inside of the balloon is the center of the universe, and the far side would be the edge of what they could see. So there's definitely a center and an edge to the universe in the balloon analogy.*

Student 2: *I think you misunderstand the analogy. The surface of the balloon is supposed to be the entire universe. The inside of the balloon isn't part of the universe and doesn't actually exist. You can't look through the inside of the balloon to the other side so there is no center in the middle or edge on the other side. In this analogy, people living in the balloon universe would never encounter a center or an edge.*

Do you agree or disagree with either or both of the students? Explain your reasoning.

9) The balloons on the previous page represent the universe at different times during its history. Draw an arrow underneath the balloons that points from the earliest time to the latest. Label the ends of the arrow with the words "earliest" and "latest."

10) Imagine you lived in a galaxy on the surface of the balloon. As the balloon expands, would all the other galaxies appear to move toward you or away from you?

11) Would your answers to the previous question be the same regardless of the galaxy in which you live, or would it change depending on the galaxy you inhabit?

12) In this analogy, do galaxies move relative to one another because they are traveling across the surface of the balloon, or do they move relative to one another because the balloon is expanding?

13) The balloon analogy is a helpful way to think about expansion, but no analogy is perfect. Some aspects of the real universe are captured by this analogy while others are not. The evidence we now have about the real universe implies the following statements (a–f) are all true. For each of these statements, state whether it is accurately captured by the balloon analogy or not, and explain your reasoning.

a) The real universe has no center.

b) The real universe has no edge.

c) The real universe is expanding.

d) The real universe is not round.

e) The real universe's expansion does not cause galaxies to change size.

f) The real universe is 4-dimensional (3 dimensions of space and 1 of time).

Part I: Expansion, Distance, and Velocity

Consider the small section of the universe containing four galaxies (A–D), shown in Figure 1 below. The distances between each galaxy are also shown.

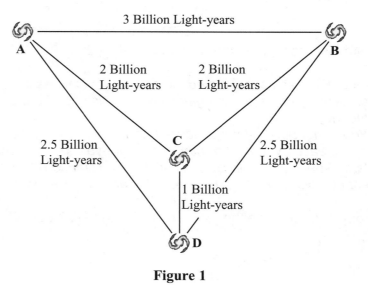

Figure 1

1) Imagine that this section of the universe doubles in size over time due to the expansion of the universe. Draw what the above section of the universe would look like after it doubles in size. Be sure to identify the new distances between the galaxies.

2) Which of the galaxies (B–D) increased its distance from Galaxy A by the greatest number of light-years during this time? Explain your reasoning.

3) Two students are discussing their answers to Question 2:

Student 1: *All of the distances doubled, so all of the distances increased by the same amount. There is no one galaxy whose distance from Galaxy A increased the most.*

Student 2: *You're right that all the distances double in size, but I don't agree that they all increase by the same number of light-years. Since Galaxy B was the farthest away from Galaxy A initially, its distance will increase by the greatest number of light-years when this section of the universe doubles in size.*

Do you agree or disagree with either or both of the students? Explain your reasoning.

4) Describe the relationship between a galaxy's distance from Galaxy A and the speed at which that galaxy appears to be moving away from Galaxy A.

5) Is the relationship you described in Question 4 unique to Galaxy A, or would you observe the same relationship (between distance and speed) if you lived in one of the other galaxies? Explain your reasoning.

Part II: Understanding Hubble's Law and Hubble Plots

The relationship you described in Questions 4 and 5 is called *Hubble's law*. We can depict Hubble's law with the graph shown at right. This graph plots the speed at which a galaxy appears to move away from us versus its distance from us. This type of graph is called a *Hubble plot*. Each dot on the plot represents a different galaxy.

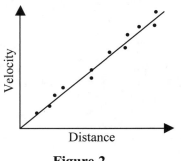

Figure 2

6) Explain how the Hubble plot shown in Figure 2 is consistent with the relationship you described in Question 4.

7) Imagine the Hubble plot shown in Figure 2 represents a universe that doubles in size over a certain amount of time. Which of the Hubble plots shown in Figures 3 and 4 below might represent a universe that triples in size over the same amount of time? Explain your reasoning.

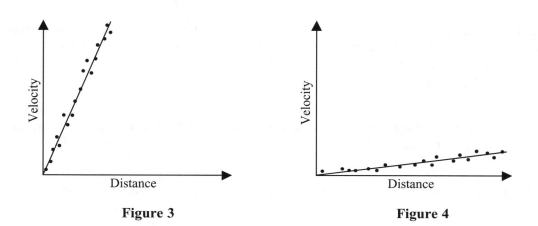

Figure 3   **Figure 4**

The *expansion rate* of the universe determines how fast the universe increases in size with time. For example, a universe that is tripling in size has a faster *expansion rate* than a universe that is doubling in size over the same amount of time. In a Hubble plot, the *expansion rate* is indicated by the slope of the graph. A steep slope indicates a fast expansion rate, while a flat slope indicates a slow expansion rate.

8) Would you say the expansion rate for the universe represented in Figure 2 is constant, increasing, or decreasing with time? Explain your reasoning.

9) Rank (from fastest to slowest) the expansion rates of the three different universes represented in Figures 2, 3, and 4. Explain your reasoning.

10) If the expansion rate of our universe had been faster, would the universe have reached its current size earlier in its history or later? Explain your reasoning.

11) If the Hubble plots in Figures 2–4 represent three universes that are the same size, which Hubble plot belongs to the youngest universe? Explain your reasoning.

12) Complete the sentence below using the words provided in parentheses ().

For two universes that are the same size, the universe with the faster expansion rate must be _____(younger/older) than the universe with the slower expansion rate. The slope of the line in the Hubble plot of the _____ (younger/older) universe will be _____ (steeper/flatter).

We can imagine many different Hubble plots, which may or may not represent how galaxies move as a result of expansion.

13) On the blank graph in Figure 5 below, draw a Hubble plot for which the expansion rate is zero.

14) On the blank graph in Figure 6 below, draw a Hubble plot for which the expansion rate increases throughout the lifetime of the universe.

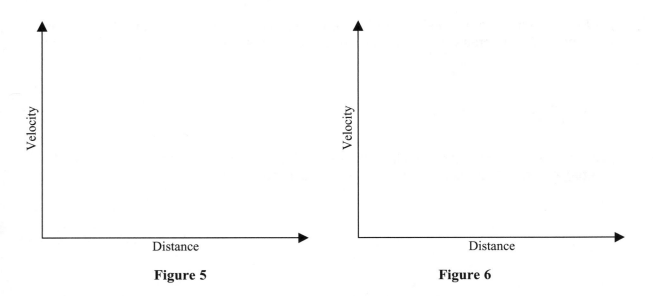

Figure 5   Figure 6

LECTURE-TUTORIALS FOR INTRODUCTORY ASTRONOMY
THIRD EDITION

Part III: Our Universe

Recent observations indicate the Hubble plot for our universe actually looks more like the plot in Figure 7.

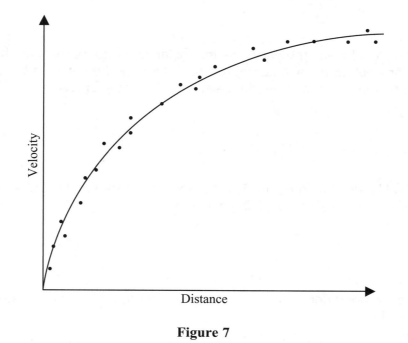

Figure 7

15) Parts a–h all refer to Figure 7. Draw or write the additional information on Figure 7 as instructed:

a) Draw a circle around the galaxies from which we receive light that was emitted closest to our present time.

b) Draw a square around the galaxies from which we receive light that was emitted furthest from our present time.

c) Write the letter C, and draw an arrow to the galaxies that are moving away from us with the fastest velocities.

d) Write the letter D, and draw an arrow to the galaxies that are moving away from us with the slowest velocities.

e) Write the letter E, and draw an arrow to the graph, where it has the steepest slope.

f) Write the letter F, and draw an arrow to the graph, where it has the flattest slope.

g) Write the letter G, and draw an arrow to the portion of the graph that corresponds to the fastest expansion rate.

h) Write the letter H, and draw an arrow to the portion of the graph that corresponds to the slowest expansion rate.

16) Based on the Hubble plot shown in Figure 7, would you say that the expansion rate of the universe is constant or changing with time? Explain your reasoning.

17) Based on the Hubble plot in Figure 7, is the expansion rate represented by the motion of galaxies far away from us faster than, slower than, or the same as the expansion rate represented by the motions of nearby galaxies? Explain your reasoning.

18) Based on the Hubble plot in Figure 7, is the expansion rate of the universe increasing or decreasing as time goes on? Explain your reasoning.

19) Consider the following debate between two students regarding their answer to the previous question:

Student 1:   *The expansion rate of our universe must be slowing down as time goes on. If you look at the Hubble plot, you can see that the graph gets flatter. That means the farther away you look, the slower the expansion rate is. The rate at which the most distant galaxies are moving away from us has started to slow down and eventually the expansion rate of nearby galaxies will also slow down.*

Student 2:   *I think you are reading the graph wrong. The slope of the graph tells you how fast the expansion rate of the universe is, not how fast a galaxy is moving. The farther we look into space, the further we are looking back in time. Since the slope of the Hubble plot is flatter in the past and steeper now, that means the expansion rate has sped up over time.*

Do you agree or disagree with either or both of the students? Explain your reasoning.

20) Based upon your previous answers, is the graph you drew in Question 14 correct or does it need to be redrawn? Explain your reasoning.

The two drawings below represent the same group of galaxies at two different points in time during the history of the universe.

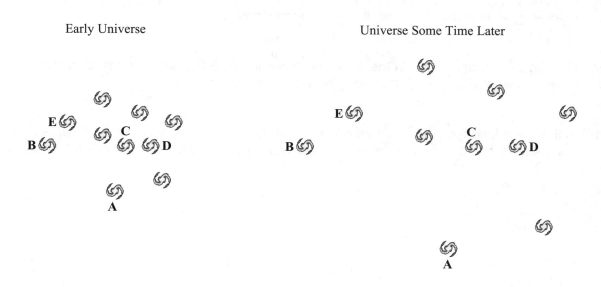

Early Universe   Universe Some Time Later

1) Examine the distance between the galaxies labeled A–E in the Early Universe. Are all the galaxies the same distance from each other?

2) Describe how the universe changed in going from the Early Universe to the Universe Some Time Later.

3) Do the galaxies appear to get bigger?

4) Based on your answer to Question 3, do you think the stars within a galaxy move away from one another due to the expansion of the universe? Explain your reasoning.

5) Compare the amount that the distance between the D and C galaxies changed in comparison to the amount that the distance between the D and E galaxies changed. Which galaxy, C or E, appears to have moved farther from D?

6) If you were in the D galaxy, how would the A, B, C, and E galaxies appear to move relative to your location?

7) If you were in the D galaxy, would the A, B, C, and E galaxies all appear to move the same distance in the time interval from the Early Universe to the Universe Some Time Later?

8) Imagine that you are still in Galaxy D. Rank the A, B, C, and E galaxies in terms of their relative speeds away from you, from fastest to slowest.

9) Now imagine that you are in the E galaxy. Rank the A, B, C, and D galaxies in terms of their relative speeds away from you, from greatest to smallest.

10) Is there a relationship between an object's distance away from you in the universe and the speed it would appear to be moving away from you? If so, describe this relationship.

11) Would your answer to Question 10 be true in general for all locations in the universe?

12) Consider the following discussion between two students regarding the possible location of the center of the universe.

Student 1: *Since all the galaxies we observe are moving away from us, we must be at the center of the universe.*

Student 2: *If you look at the drawing on the first page, it's pretty clear that all the galaxies move away from each other, so I think each galaxy must be at the center of the universe.*

Do you agree or disagree with either or both of the students? Explain your reasoning.

When the universe was 4 billion years old, Galaxy A was 3 billion light-years away from Galaxy B, as shown below. Imagine that the universe was not expanding, so the distance between Galaxy A and Galaxy B would not change over time.

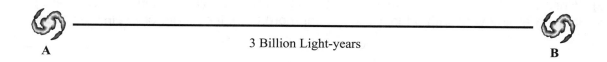

3 Billion Light-years

1) A star explodes in Galaxy B producing a large amount of light. How long will the light from this explosion take to reach Galaxy A?

2) How far did the light travel on its journey to Galaxy A?

3) How old will the universe be by the time the light from the explosion reaches Galaxy A?

Because light takes time to travel from place to place in the universe, when we look at the night sky we are seeing stars and galaxies as they appeared in the past. For example, if we see a galaxy 1 million light-years away, we are seeing what the galaxy looked like 1 million years ago. We say this galaxy has a lookback time of 1 million years. Lookback time is the amount of time light takes to travel to us from a distant object.

4) When inhabitants of Galaxy A see the light from the explosion, what is the lookback time they associate with Galaxy B?

Our universe is expanding. This means the distance between galaxies is constantly increasing. Imagine that Galaxy A and Galaxy B are in an expanding universe.

5) While the light from the explosion is traveling from Galaxy B to Galaxy A, does the distance between the two galaxies stay the same, become larger, or become smaller?

6) By the time the light from the explosion in Galaxy B reaches Galaxy A, is the distance between the galaxies more than, less than, or exactly 3 billion light-years?

7) By the time the light from the explosion in Galaxy B reaches Galaxy A, has more than, less than, or exactly 3 billion years elapsed since the star exploded?

8) By the time the light from the explosion in Galaxy B reaches Galaxy A, will the total distance traveled by the light be more than, less than, or exactly 3 billion light-years?

9) When the inhabitants of Galaxy A see the light from the explosion in Galaxy B, are they looking at an event with a lookback time of more than, less than, or exactly 3 billion years?

10) In the space below, provide a sketch that explains the reasoning behind your answers to Questions 5–9.

11) Consider the discussion between two students regarding their ideas about two distant galaxies in an expanding universe.

Student 1:   *Let's say light takes 5 billion years to travel from one galaxy to another. This means the two galaxies were separated by 5 billion light-years when the light began its journey.*

Student 2:   *If the light traveled for 5 billion years, then the distance between the two galaxies must have been less than 5 billion light-years when the light began its journey because the distances between galaxies are always increasing in the expanding universe.*

Do you agree or disagree with either or both of the students? Explain your reasoning.

Diagrams A and B below each represent a different way of thinking about how very large regions of the universe change over time. The dots in each diagram represent pieces of matter.

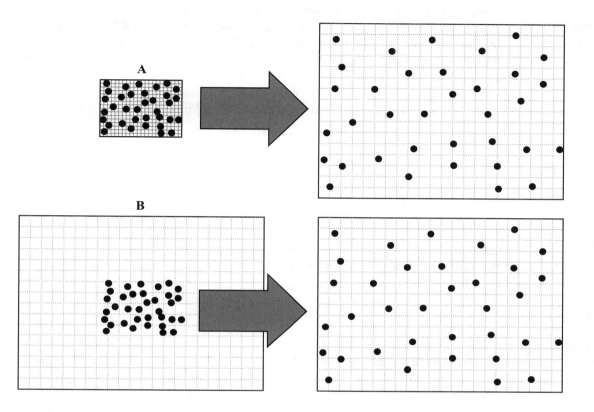

1) Which diagram, A or B, is a better representation of the universe we observe? Explain your reasoning.

2) In Diagram A, is the universe becoming bigger, smaller, or staying the same size over time?

3) In Diagram B, is the universe becoming bigger, smaller, or staying the same size over time?

4) Two students are debating their answers to Questions 2 and 3:

Student 1:   *Both diagrams show the universe becoming bigger. In Diagram A, the grid has expanded and become larger. In Diagram B, the pieces of matter have spread out and take up a greater amount of space.*

Student 2:   *I disagree. Only Diagram A shows the universe becoming bigger. In Diagram B the size of the grid doesn't change. The pieces of matter are just moving into an already existing empty space in a universe whose size doesn't change.*

Do you agree or disagree with either or both of the students? Explain your reasoning.

5) Both diagrams show the distance between matter increasing over time.
   a) Which of the diagrams shows this happening as the result of space expanding and which is a result of an outward explosion?

   b) Which of the diagrams is a more correct representation of our universe? Is your answer to this question consistent with your answer to Question 1? Explain your reasoning.

Consider the three diagrams (C, D, and E) shown below. These diagram each represent a single region of the universe, but at different times during the history of the universe.

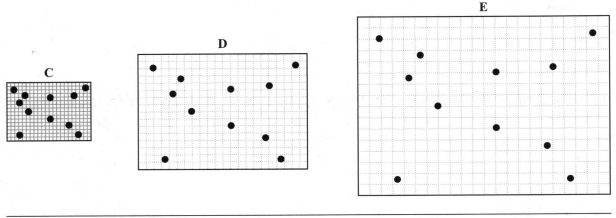

6) Draw an arrow below Diagrams C, D, and E. The arrow should point from the diagram that represents the earliest time in the universe's history to the diagram that represents the latest time in the universe's history. Label the ends of the arrow with the words "earliest" and "latest."

7) In which drawing does the region of space have:
   a) the highest density?

   b) the greatest concentration of energy?

   c) the highest temperature?

   Explain your reasoning.

8) Imagine you could watch the history of the universe like a movie <u>playing backward</u>. The movie starts today and ends at the beginning of the universe. Describe what you would see for every region of the universe as the movie played and you looked further back in time. Your answer should discuss how regions of the universe change in terms of temperature, density, and size.

Your answers to the previous questions are all part of the *Big Bang theory*. The Big Bang theory describes the universe as starting 13.7 billion years ago and how it has changed over time.

9) Three students are discussing their understandings of the Big Bang theory:

Student 1: *I think I understand the Big Bang now. At the beginning, all the matter in the universe was compacted into a small, hot, dense ball. This ball of matter then exploded into empty space. When we look at the universe, we see galaxies moving away from us. The Big Bang model explains this, since all matter should be flying away from the center point of the explosion.*

Student 2: *I disagree. I think what the Big Bang theory is saying is that all the matter in the universe was once compacted into a really dense and hot object that expanded over time. But there wasn't an explosion of matter into empty space. Instead, the universe carried pieces of matter away from each other as it expanded in size.*

Student 3: *You're both wrong. I agree that the universe was once smaller in size and that pieces of matter have been carried away from each other by the expansion of the universe. But remember how we learned from Einstein's equation $E = mc^2$ that matter can be converted into energy and energy can be converted into matter? I think this means that if we go back to the beginning of the universe, it would be so incredibly dense and hot that matter itself couldn't exist. I bet at the very beginning, the universe would have been composed of pure energy with no matter there at all.*

Do you agree or disagree with any or all of the students? Explain your reasoning.

10) Based on your previous answers, complete the following sentences:

The Big Bang theory says that the universe started out with a/an _____ temperature and a/an _____ density. Originally, there was no _____, only pure _____. From this initial state, each region of the universe _____ in size. This caused its temperature and density to _____. When the temperature was cool enough, energy could transform into _____.

11) Look at Diagram A again. Next to Diagram A, make a drawing of what you think that region of the universe would have looked like at the very first instant it existed.